A VISUAL DIRECTORY OF 100 OF
THE MOST POPULAR SONGBIRDS

A FIELD GUIDE TO
SONGBIRDS
OF NORTH AMERICA

NOBLE S. PROCTOR PH.D

chartwell
books

Inspiring | Educating | Creating | Entertaining

Copyright © 2016 Quarto Publishing plc

This edition published in 2021 by Chartwell Books, an imprint of The Quarto Group
142 West 36th Street, 4th Floor, New York, New York 10018 USA
T (212) 779-4972 F (212) 779-6058
www.QuartoKnows.com

ISBN: 978-0-7858-3947-7

Library of Congress Control Number: 2020952760

QUAR.SBG

Conceived, edited, and designed, by
Quarto Publishing, an imprint of The Quarto Group
The Old Brewery
6 Blundell Street
London N7 9BH

Project editor: Liz Pasfield
Art editor/Designer: Tania Field
Assistant art director: Penny Cobb
Copy editor: Claire Waite Brown
Bird artist: David Ord Kerr
Illustrators: Vana Haggerty and Wayne ford
Picture researcher: Claudia Tate
Art director: Moira Clinch
Publisher: Samantha Warrington

Printed in China

10 9 7 8 6 5 4 3 2 1

The material in this book was previously published in
The Songbirds Bible.

A FIELD GUIDE TO
SONGBIRDS
OF NORTH AMERICA

Contents

Why birds sing

When listening to a bird's song, it is often tempting to interpret its reasons for singing in solely human terms. Birdsong gives us pleasure, but from the birds' point of view, it is an entirely functional form of communication. Singing is directed at members of a bird's own species, and it follows that each species has a different song, while individuals of each species sound alike. However, there is an occasional variation between different populations of the same species, in the form of dialects, that can distinguish between subspecies.

In general, it is the male that sings, although the females of some species can also sing, though usually not as well. Males use song to draw attention to themselves, whereas females in the vulnerable position of incubating eggs or young want to remain hidden.

ESTABLISHING HOME BOUNDARIES

One of the prime reasons for birdsong is to establish the existence and boundaries of a territory. It is within this territory that a pair of birds will raise their young, so it must be jealously guarded from rivals as it will provide food and protection for both parents and young until the breeding season is complete. Singing lets other males of the same species know that a family is in residence and that intruders are not welcome. The song means nothing to other species that live in the territory, as these will often have different food and nesting requirements anyway. A singer on its territory will ignore their presence, just as they in turn ignore his. Within a given territory, it is possible to have robin, cardinal, catbird, and yellowthroat all nesting without interfering with each other's needs.

The male chooses a series of song posts from which to sing and by singing from each of them regularly, he can define his territory. The borders are invisible, of

Both the male and female wood thrush care for the young, but the male still sings to protect the territory.

course, but if ever a rival male crosses into an occupied territory, he will be instantly challenged by the occupant. The ensuing fights or chases may sometimes involve aggressive bursts of song to help see off the intruder. The territory's original "owner" invariably emerges as the winner of such disputes. Some cunning birds have an entire repertoire of songs, and use a different one at each song post. This may well create the illusion of a number of occupied territories and be successful at keeping would-be intruders away.

EFFECTIVE ADVERTISING

Another use for birdsong is to advertise to a female that there is an unattached male present who is offering her a home. It follows, then, that a male that is paired up will tend to sing less, using his song purely for territorial reasons, though how a female can distinguish between the songs of a bachelor bird and one that is happily paired is not clear. Singing also serves to strengthen the bond between a pair and may also be involved with further courtship leading to mating.

Some birds may also have a subsong, often a quieter and lower version of their normal song. It does not carry for any distance and is usually heard only in fall and early spring. Some subsongs are almost certainly rendered by young birds that have not yet reached the stage of full

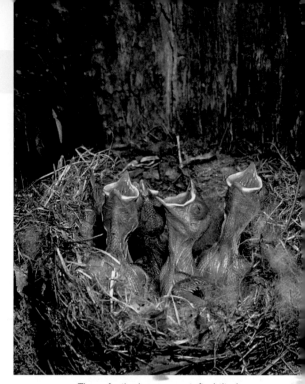

These featherless young tufted titmice start learning the song of adults well before fledging.

territorial singing, and could therefore be regarded as a form of practicing.

Another form of song is a type of excited outburst that happens at random and contains many improvised elements, unlike the stereotyped territorial song. The outburst may be delivered from a perched position, but is more often given in an "ecstasy flight." It has been suggested that this sort of song may be a form of emotional release. The outburst usually occurs at the peak of the breeding season, often at night or twilight. Mockingbirds, yellowthroats, and ovenbirds are some species that regularly indulge.

Bird calls are functionally different from songs, and so have a different structure. They are short simple sounds, usually of one or two syllables. Some are more complex, however, and almost fall into the song category, just as some songs are simple and could be mistaken — by the human ear, at least — for calls. Calls, however, communicate totally different messages from songs.

A VARIETY OF CALLS

Songbirds tend to have a larger repertoire of calls than other birds, many having a vocabulary of 20 or more. Calls may be used to communicate with a partner, to beg for food, to call young birds, to keep contact with flock members, to show aggression, or to signal that a predator is near.

Once the male's song has attracted a mate, calling between pairs often forms part of the courtship ritual. Courtship feeding by the male is usually accompanied by a begging call from the female. Parent birds can call young birds to them when they have scattered after leaving the nest, particularly useful for precocial species (mobile before they are fully grown). Migrating birds call to one another when flying in a flock, a practice that enables a nocturnal migrant to rejoin the flock if it becomes separated. Feeding flocks in woodland call constantly; this helps them to forage more effectively and enables them to signal when they locate a food source. Winter flocks of birds like

Some species such as mourning doves feed in groups and then return to the nesting territory to call and keep other doves away.

This northern cardinal views its own image in a car mirror as a territory aggressor and will try to fight it for hours.

chickadees have foraging territories that may be defended from other flocks. This is particularly important in times of food shortage, when each member of the flock will identify themselves by a call. If an individual who doesn't belong to the flock — and who therefore has a different call — enters the foraging area, it is immediately spotted and driven away.

Perhaps the most interesting type of call is the one that alerts birds to the presence of a predator. The alarm call has the same basic pattern for a wide range of species. It is a short, high-pitched note that can be heard clearly, but gives little information about the location of the calling bird. This call usually results in all of the small birds taking cover and is given when a bird of prey or predator is spotted nearby.

A different type of alarm call will be sounded if a stationary predator is seen; owls usually provoke this type of call if they are discovered roosting. Instead of a short, high note, a loud scolding type of call, repeated many times, draws attention. This sound is interpreted by other species who usually come to investigate before joining in. The resulting chorus of scolding noises is part of a mobbing display, and the birds will often approach very close to the object of their attention. If the owl moves, the short, high call is given and all dive for cover. Mobbing usually persuades the target to move on.

How birds sing

The vocal sounds produced by birds are made in a completely different manner than those produced by mammals. Instead of having a larynx with vocal cords situated at the top of the trachea, birds possess an organ called the syrinx. This is a V-shaped structure situated at the base of the trachea (windpipe) where it divides into two bronchi that run to the lungs. Inside the syrinx are thin tympanic membranes that vibrate when air passes over them as it escapes from the lungs. The shape of the syrinx can be altered by muscles attached to it. This in turn changes the shape and tension of the tympanic membranes, which then varies the pitch of the sound produced. The more muscles to control the syrinx, the richer the range of sound. Songbirds have from five to nine, which give them their wide vocal abilities.

VARIATIONS ON A THEME

Of songbird species, ovenbirds, wood creepers, and antbirds (the suboscines) have the most primitive type of syrinx, with the membranes attached only to the trachea. The rest of the songbirds have membranes that can be attached to both the trachea and bronchi in different ways.

Part of a bird's respiratory system is formed by a series of air sacs; these allow a greater volume of air to be taken in than the lungs alone would hold. The

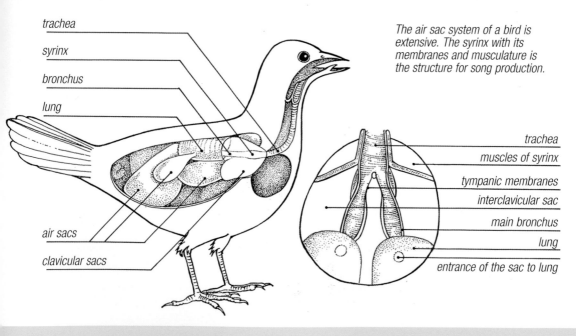

trachea
syrinx
bronchus
lung
air sacs
clavicular sacs

The air sac system of a bird is extensive. The syrinx with its membranes and musculature is the structure for song production.

trachea
muscles of syrinx
tympanic membranes
interclavicular sac
main bronchus
lung
entrance of the sac to lung

interclavicular air sac surrounds the syrinx and exerts pressure on it. This particular air sac is essential for the production of sounds; if ruptured, no sounds can be made. The trachea also plays a role by acting as a resonance chamber.

Among the different bird species, there is a great variation in syrinx structure and, therefore, the types of sound produced. The length and width of the trachea also play a part in the final sound. A short, narrow trachea produces a higher resonance than a short, broad trachea. As song originates from the base of the trachea, it is possible for birds to sing with their bills full of food, or even closed.

Unusual sounds

Some species, not noted for their songs, use their air sacs to produce extraordinary sounds. Greater and lesser prairie chickens inflate theirs to produce a booming sound, and the sacs can be seen stretching the bare neck skin as they are inflated. Pigeons inflate their esophagus with air to give their characteristic cooing tone. The modification of syrinx, trachea, and air sacs is what makes birdsong so beautifully varied and unique.

Many birds — though relatively few songbirds — produce sounds other than vocal noises by using parts of their bodies. Hummingbirds are so-called because of the noise their wings make, and the varying species each produce a different hum. Other parts of the body can be used to produce sounds. Storks clap their bills together to make a loud rattling sound, and woodpeckers drum with their bills against a tree or branch.

Mimicry

There are some species where the majority of the song appears to be innate. Song sparrows, for instance, do not need to hear a parent or any other song sparrow to develop a perfect song pattern. When they do hear one, however, they develop a song that mimics the song sparrow dialect they hear. As a result, song sparrows may have as many as 50 distinct local dialects.

As well as learning their own songs, many birds are excellent mimics, building the phrases of other species into their own songs. The Indian hill mynah is an accomplished mimic in captivity, to the extent that it can even copy human speech; in the wild, however, it does not incorporate the songs of other birds into its own. The marsh warbler breeds in Europe and winters in Africa, and its song is made up almost entirely of phrases from other species. Studies have shown that these consist of a mixture of phrases from nearly 100 European species and over 100 African species. The most notable North American mimic is the mockingbird, known to imitate 55 species in an hour.

When birds sing

In temperate regions, the amount of daylight plays an important part in the life of all birds. It tells migrants when to migrate, and it also triggers the production of hormones that prepare birds for the breeding season. Singing is stimulated by the presence of the male sex hormone, testosterone, produced when

In some species, the male not only sings to the female but also displays to her at the same time.

daylight length reaches a certain amount — usually in excess of 12 hours. In temperate regions, this results in a distinct breeding season during which the majority of birds breed. In the tropics, where the change in day length is minimal, other factors may also be responsible, which is why breeding occurs throughout the year in such areas. As birdsong is linked so closely to breeding, most singing takes place during the breeding season.

The length of song period varies from species to species, and the timing of the breeding season differs geographically. Birds that breed in temperate regions have a clearly defined breeding season. Many birds begin singing in earnest at the beginning of the year, so as to establish their territories early on. As spring approaches, more and more birds sing, and with the arrival of spring migrants the chorus is complete. In the northern hemisphere, there is nothing to compare with the woodland dawn chorus in early May.

For the most part, spring visitors will have been silent on their wintering grounds, but as they prepare to migrate they often begin to sing. The song begins very tentatively to start with, and far from perfect; many young birds will be singing for the first time. When these migrants eventually reach their breeding grounds, they will be in full song, ready to establish a territory and find a mate.

PROMPTED BY DAYLIGHT

Lengthening daylight hours encourage resident birds to sing; as the days get longer their singing becomes more pronounced and continues for longer periods. Wintering birds are often silent for the whole of their stay south, but occasionally some departing late may be stimulated into song by a particularly

Red-wing blackbirds sound out from the singing perches at dawn and dusk, and forage for food at midday.

springlike day. Birds can be fooled by light levels. In addition, some will begin to sing in fall when the daylight length matches that of spring. This singing soon stops as the days get shorter, however.

In most species, song is at its fullest at the start of breeding, with males singing throughout the day. It wanes slightly during courtship and mating, and picks up again during the incubation period, although males that help with the incubation will sing less. As soon as the eggs hatch and help is needed to feed the young, the male will sing less often, concentrating his territorial song on the area around the nest.

Many birds go silent and molt when their young fledge and need no more

attention. Others may have a second brood; if so, the male will sing again with renewed ardor, often re-establishing territorial boundaries. In late summer or fall the hormonal drive tails off, and many birds stop singing altogether, though this pattern of song is a generalization, and there are some species that do not conform. The mockingbird and the cardinal, for instance, sing in every season as they defend their territories throughout the year, while the brown thrasher ceases singing immediately after mating.

As well as this annual variation in song, there is also a daily rhythm. Birds that are active during the day will begin singing just before sunrise as the increasing light reaches a certain intensity. As the morning progresses, birds begin to quiet down; singing picks up in the late afternoon, continuing until nearly dusk.

Different species are triggered into song by varying amounts of light, and when listening to a dawn chorus, separate species can be noted as they start. The order in which birds first sing is fairly constant, though it depends upon the birds present. If they are in the vicinity, robins are usually the first to pipe up. In eastern states, the next to sing might be a wood thrush and then a Carolina wren, while out West it would most likely be a hermit thrush and then a Bewick's wren. Chickadees, phoebes, towhees, and sparrows also start singing early.

It is interesting to note that thrushes are commonly found to be the first singers in most countries. In Europe, the blackbird is often first, followed by the song thrush. In India, one of the first is the Indian robin; in southern Africa, it is the kurrichane and olive thrushes. In New Zealand, with its high number of introduced European birds, it is once again the blackbird and the song thrush that begin the chorus.

The contrast between the dawn chorus and the quiet of midday is truly astonishing when you realize the birds are still wide awake, though once again there are exceptions, as certain birds sing consistently throughout the day. Field sparrows, indigo buntings, red-eyed vireos, and prairie warblers are just a few

The dawn chorus can be overwhelming with birdsong. Here an eastern meadowlark adds its song.

Some species enhance their vocal ability with elaborate displays. Early literature often depicts the displays of species such as lyrebirds (top), redstarts (bottom left), and birds of paradise (bottom right).

examples. Actually, the red-eyed vireo holds the world record for the number of songs given in a day — a total of 22,197!

Nighttime vocalists

Night is not solely reserved for owls — it is often surprising just what else can be heard under the cover of darkness on a spring night. Mockingbirds are the North American equivalent of the legendary nightingale and can be heard regularly at night, together with yellow-breasted chats. Outbursts of "ecstasy" singing from yellowthroats and ovenbirds add to the night chorus that, of course, also includes species such as whippoorwills.

Where birds sing

The tree pipit delivers its song while floating down to the ground with its wings outstretched.

As the main purpose of song is to deter competitors and attract a mate, it follows that the further a song carries, the better. For this reason, many male singers will choose a prominent perch, such as the top of a tall tree, bush, or rock, and sometimes an artificial song post at the top of a telephone pole or tall building. The song post may be positioned at the edge of a territory, whose boundaries are often marked by a number of these song posts. The actual nest is normally sited well within the territory and, especially with large territories, the male will sing from a perch well away from the nest so as not to attract predators directly to it.

For bird watchers, the fact that different singing birds choose different but consistent perching places helps to locate the bird. This is particularly true of many woodland species, especially warblers. When spring migration is in full swing, a piece of woodland can contain dozens of different species, and it helps enormously that black-throated blue warblers sing from the undergrowth, black-throated green warblers from a middle height, and Cape May warblers from the very top. These levels are also the birds' usual feeding points, so they can be looked for with a greater chance of success even when silent. However, singing, feeding, and nesting levels are not always the same; many thrushes that feed on the ground, for instance, will nest off the ground and sing from the top of a tree. Also, in open areas such as fields, prairie, tundra, and desert, many birds deliver their songs during flight. Larks, longspurs, and pipits will "perch" in the air to deliver their songs.

Birds that sing from exposed perches tend to have shorter song phrases, possibly to reduce the chance of a predator catching them unawares in midsong. Certainly some of the longest songs come from species that stay well hidden, such as the grasshopper warbler, which has a trilling song lasting for more than two minutes. The possible extra vulnerability means that normally song perches are positioned where the singer can see a predator coming, and also near to cover for escape. When birds are establishing

song posts, they often try out a few,
eventually settling for the safest one
that lets them communicate to the
widest audience.

*A male rufous-sided towhee
announces his territory from
a singing perch.*

Identifying songbirds

To the inexperienced bird watcher, the way in which a practiced ear can pick out and identify bird sounds seems almost supernatural. Most beginners cannot imagine ever mastering the different sounds, but with practice, it becomes relatively easy. Most people aware of birds will recognize the songs of many birds around them. Learning to recognize the commoner species provides a basis from which to expand. Having learned the songs and calls of birds in your backyard, visit a woodland where many of these familiar birds will be present. Whenever possible, go out with a more experienced bird watcher who will be able to draw your attention to the similarities and differences of songs. An additional help may be a recording, so that it can be played again and again, thereby familiarizing yourself with the birdsongs.

TRANSLATING BIRDSONG

Some birdsongs are easier to remember than others, since they lend themselves to verbal descriptions. If you aren't able to record the birdsong, it is useful to familiarize yourself with a means of writing down songs and calls, as this may help to identify them later. Written descriptions are possibly the best way to do this, though what you hear and write down may not correspond with what is written in a book. It is worth practicing by writing out songs from a recording and comparing them with the book.

RECORDING AND VIEWING

Many bird enthusiasts take up bird sound recording. To get the best results, specialized and often expensive equipment is needed. Nowadays, large tape recorders of the past have all but vanished and portable cassette recorders and light, hand-held microphones can produce high-quality recordings. With present-day technology, it is also possible to use your smart phone. If you buy cheaply, the results are likely to be poor, so it's always worth investing in quality equipment. There is such a wide choice of microphones and recorder, plus a bewildering number of terms to contend with, that the best way to choose is to seek advice first from a bird watcher who already records sounds.

For viewing, the barest of essentials are binoculars and a good field guide. On the market today, there is an overwhelming selection of binoculars. Which one you end up with is often dictated by your budget — but a few things should be considered. The very expensive models are optically perfect for viewing under a wide range of conditions. Grinding of the glass allows for maximum use of light under all conditions. They are usually waterproof and dustproof. The prisms of the expensive models are anchored in place

One of the best ways to learn about birds is to go out with groups and learn from the experts.

much better, and this prevents slippage in heat or when jarred and thereby prevents knocking the binoculars out of alignment. When you pay a lower price, many of these features are lost. In addition, the more expensive glasses can focus extremely close, a feature that is a must for tropical or dense thicket birding. However, for $250 or so, a good pair of binoculars can be had that will fit the average birder's needs. As for power, ask three people and you might get three answers. The standard, and a good choice, is 7 x 35 or 8 x 42 with center and right eye focus. Seven or eight power will allow a very wide depth of field and eliminate the need to keep focusing and refocusing.

Do not be fooled into "the more power, the better I'll see the bird" syndrome. People afield with 15 and 20 power binoculars will find them worthless.

Take the time and look through the binoculars you will purchase; they need to be right to make birding fully enjoyable.

OUT IN THE FIELD

Familiarity with songbirds is gained by many hours afield; hearing unfamiliar calls, tracking the birds down, and then studying the individual as it sings. As this is being done, many other fragments that will lead to birding competence are being collected. You will begin to understand habitat preference, activity hours for a specific species, preferred levels of activity within the forest for feeding and singing, time of the year when the bird is most active, and perhaps information on the bird's food preference.

Backyard birdwatching

Proper plantings such as sunflowers, cosmos, and thistle are sure to bring the American goldfinch to the backyard.

a new visitor to their backyard as they do from sighting the rarest of vagrants.

For a yard to be attractive to birds, it must provide sufficient food, water, breeding sites, and protection. The most successful ones will reproduce something of the birds' natural habitat, especially those in which some parts have been left in a wild state to provide seeds and berries for food, and thick cover for nesting and roosting. It should also be left untreated by pesticides in order that a full community of insects and invertebrates can also live there and provide further food. Such an area should have something to attract both breeding birds and winter visitors, ensuring a variety of birds throughout the year.

PLANTING FOR SUCCESS

The most attractively managed gardens for birds will contain: food plants, which can be a mixture of native and ornamental types; trees and shrubs, to provide further food, shelter, nesting sites, song posts, and protection from predators; nest boxes, to provide suitable sites that do not occur naturally; a source of water, both for drinking and bathing (either a pond or a simple birdbath); and a winter feeding

For those people who enjoy the natural world around them, the presence of birds in their own backyard or garden can only bring pleasure. Enthusiasts lucky enough to have large gardens can entice a wide variety of species by providing the right environment, and even small yards can be made attractive to birds. Many bird watchers get as much enjoyment out of

Conifers and evergreens protect nests well and are very useful for roosting birds in the winter.

or junipers also provide berries, and white pine forms cones that hold seeds for siskins, nuthatches, and many other species. Berry-bearing bushes and shrubs are also essential — rose, hawthorn, holly, buckthorn, and elderberry are some of the best. Of the non-native shrubs,

Fruiting trees, such as mountain ash and European rowan, provide food for many species, such as waxwings, during the winter months.

station, to provide a variety of foods (and that is visible from a watch point).

The food plants, trees, and shrubs that are best for a bird garden will depend on what region of the country the garden is in, as well as its size. Native trees providing both cover and food are ideal. Red oaks provide nesting places and cover, and produce acorns; they grow best in the East. Beech produces a nut which is eaten by woodpeckers and blue jays; ash has a winged seed eaten by purple finches and grosbeaks; and mountain ash berries attract waxwings, robins, and orioles. Fruits from various types of cherry are eaten by more than 80 species. Evergreens provide some of the best protection, while red cedars

cotoneaster produces a prolific harvest of berries and comes in a wide variety of cultivated forms. Climbing plants such as ivy, honeysuckle, and Virginia creeper are useful in smaller yards, where they can grow up a wall to provide nesting and roosting places, as well as berries.

Birds that prefer low scrub and bushes can be enticed into a garden if a brush pile is made: a thick tangle of pruned branches and swept leaves should be left in a corner where birds can hide themselves. The shyer catbirds, thrashers, towhees, and thrushes especially like these arrangements. Also, in winter, the branches can harbor invertebrates for wrens and other insect-feeders to find.

Many attractive flowers produce seeds for winter finches. Sunflower heads can be left on the plant, or collected and put out at a feeding station. Their seeds are great favorites of all birds, and are eaten by chickadees, titmice, nuthatches, cardinals, grosbeaks, sparrows, and finches. Also, thistles can be allowed to grow in a more secluded area to attract goldfinches, redpolls, pine siskins, and other finches.

FEEDERS

The provision of feeding stations during the cold winter weather will attract an even wider variety of birds to your garden. In smaller yards that don't have enough space to plant all the different trees and

Instead of buying an expensive feeder, try using a suspended jar that keeps the food dry and attracts titmice.

An empty coconut shell can be filled with a "birdcake," made from melted fat and seeds, that has set before hanging.

A tree stump can be used as a feeding station by boring a hole in it and filling it with fat, seeds, fruit, and nuts.

A seed hopper will gradually dispense seed into the tray and allow many days' supply to be put out at one time. It is especially useful if someone is going away.

A mesh bag filled with peanuts attracts finches. Hanging it from a long string stops larger birds from feeding.

Peanuts in their shells can be threaded and hung out. It is fun to watch chickadees open the shells to get at the nuts.

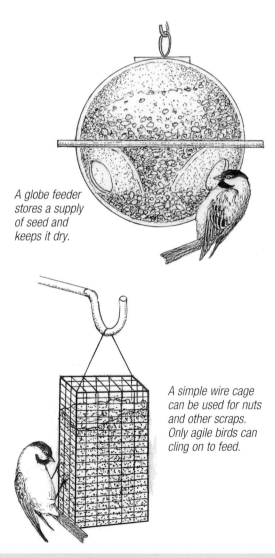

A globe feeder stores a supply of seed and keeps it dry.

A simple wire cage can be used for nuts and other scraps. Only agile birds can cling on to feed.

bushes that birds enjoy, feeders can do almost as well, giving them food and bringing you pleasure. However, keep in mind that a large number of birds may attract predators that are after a feathered meal. Always make sure, therefore, that any feeding area has some safe cover nearby that birds can fly to for protection. Ideally, feeding stations should be situated out in the open where any birds can see a potential enemy approaching. If it is too close to a bush, a cat might hide there. But the garden itself away from the feeder should be planted with dense bushes and shrubs, especially evergreens, which provide protection from avian predators such as hawks.

Many birds forage over a large area for their food, and even birds that visit feeders may go to several gardens in turn. As a result, it is difficult to judge how many birds actually make use of the garden. You may see no more than a dozen chickadees at a time, but there

A small log can be drilled with holes and stuffed with suet and nuts. Woodpeckers will feed from it.

An elaborate feeding station can consist of a roofed platform with seed hopper and nut feeder. The platform has slots at each corner to allow rain to drain away and a squirrel baffle on the post.

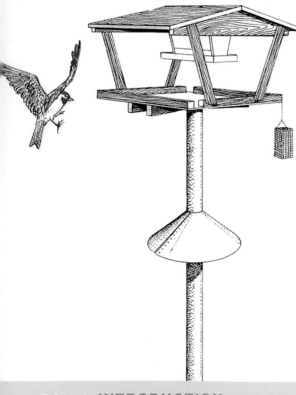

could be more than 50 visiting each day. Their foraging instinct may be such that they tend not to feed in one place for long before moving on. Then again, some species, like evening grosbeaks and house finches, will feed as long as there is food; they will rest near the feeders when the food is finished and feed again when more is put out.

A simple food platform — a wooden tray fixed onto a post, window sill, or tree stump — will bring many birds right up to your window. A roof is not essential, but it does serve to keep food dry. Feeders can be hung from the platform or from tree branches, and these can be used to dispense seeds, nuts, and other food.

Variety of food

Ultimately, the key to a successful feeding station is to place out as wide a variety of foods as possible. Seed mixtures contain different-sized seeds for different birds, for instance, millet, thistle, hemp, sunflower, corn, and wheat. Peanuts and peanut butter are very popular; the latter can be smeared onto branches or placed out in log feeders. Fatty foods like suet provide birds with the most energy. It can be placed out in lumps, or melted and mixed with seeds to form a cake. Scraps of food such as cheese, meat fat, and bones will also be gratefully received. Fruit is a favorite of many birds — apples, raisins, cherries,

The squirrel is the bane of many feeders and can test the skill and patience of the backyard birder.

and oranges will all be eaten, and are excellent as summer food.

Bird feeders are used most after the breeding season, in fall around October through winter until early spring. Remember that while food attracts birds to where you can see them, it is only in severe winters that feeding may be vital. Also, feeding throughout the year is not always beneficial for birds. Many will leave the feeders as soon as natural foods become available again in spring, but some will continue to use them if food is present. Birds change their diets as the natural food sources change; those which feed on nothing but nuts and seeds in winter will be eating insects and grubs in summer. The young of these birds need soft insect food when they are developing, however, and if the parents give them unnatural foods from a feeder, it can cause harm. Young birds may be unable to digest it properly, and it might not contain all the necessary nutrition for their normal growth.

Attracting hummingbirds

One group of birds that can be readily fed in the summer are the hummingbirds. Although not songbirds, they still make an attractive addition to the garden, as do the flowers they feed on, such as trumpet vine, fuchsia, sage, columbine, nasturtium, petunia, impatiens, and morning glory. Artificial feeders with a sugary solution inside can also be used, but care must be taken to make up the solution correctly and to change it regularly. One part sugar mixed with four parts water, boiled and cooled, produces a good energy-rich food. It must be stored in a refrigerator, and feeders should be cleaned before refilling.

With large numbers of birds feeding, a considerable amount of droppings will accumulate around the feeding stations, which should be moved occasionally to avoid contaminating the food of ground-feeding birds. In addition, the accumulation of droppings can facilitate the passing on of diseases from one bird to another. If the feeding stations can't easily be moved, it's best to rake the area, turn over the soil, and spray on a 10 percent solution of water and bleach.

Backyard predators

Cats and squirrels can be a great nuisance in a good bird garden. The former scare away the birds and occasionally catch them, while the latter will steal food from feeding stations and may also attack nest boxes. Feeding stations and nest boxes on posts can be protected by using a baffle halfway up the post. Barbed wire is unsightly but often effective, as well, although thorny cuttings from roses, holly, or hawthorn, tied in a bundle, work just as well. A bell-collar on a cat may give the birds warning of its presence and prevent them being caught. If a cat has a favorite position on a fence or under a bush, some prickly cuttings placed there will soon convince it to move on.

Some birds store the food they find. Nuthatches, titmice, and jays, in particular, will frequently take seeds, nuts, and suet that they then hide in a crevice or in the ground. They will return later for it — if it has not been stolen by other opportunist birds or squirrels that have seen them hide it!

Winter months

It is debatable whether putting food out for birds during winter actually helps them to survive, except in the severest conditions. During a mild winter, the supply of natural food available will often keep birds away from feeders, especially if the trees have produced a good nut crop. The time when birds most need food is when the temperatures are low and snow covers their normal food sources. Food is all-important to birds as it provides the energy they need to survive. At night, they lose energy in the form of heat, and during the day, they lose more in the search for food. If they can't find enough food to replace the energy they have lost, they will eventually die. It is not just the provision of food that helps but also the fact that it is easily found and is there when needed. Food should always be placed out regularly during winter months so that the birds can anticipate it and, if need be, rely upon it.

A good bird garden will also provide roosting places for birds so they don't need to travel far in order to feed when the weather is at its coldest. Sparrows, grosbeaks, and finches will simply perch deep within an evergreen, sheltered from the wind, rain, or snow. A yellow-rumped warbler has even been known to roost each night on a Christmas tree in someone's house! Some birds will make use of a vacant nest box; there is one case recorded of 31 winter wrens using the same box, and another of 12 bluebirds huddled in a box overnight.

A source of water is an important attraction as this bathing female scarlet tanager shows.

During winter months, water is as important to birds as food. They not only need to drink water, but must have it to bathe in, as well. Birds' feathers are very important as insulation and to perform efficiently, must be clean and dry. The air trapped underneath the feathers helps to insulate, which is why birds fluff out their feathers in cold weather. Regular bathing and preening enables them to clean and waterproof their feathers with natural oils. A supply of clean water should be available at all times and should be kept ice-free. Water is also equally important throughout the summer, when fresh water may be difficult to find. A simple birdbath will suffice for most birds, but a small pond with suitable vegetation near it will often attract birds that are otherwise difficult to spot in the wild. Never make a pond too deep, and be sure there are shallow edges. Branches laid into the water will provide perches.

FEEDING REQUIREMENTS

SPECIES	FOOD/FEEDING
GREAT-CRESTED FLYCATHER	Insects and fruit. Feeds among mature trees in orchards.
TREE SWALLOW	Insects caught over ponds and rivers; bayberries in fall.
BLUE JAY	Omnivorous, especially nuts and fruit, takes sunflower seeds, peanuts, and suet at feeder.
BLACK-CAPPED CHICKADEE	Insects, seeds, and fruit; takes peanuts, cornmeal, and sunflower seeds from feeder.
TUFTED/OAK TITMOUSE	Seeds, nuts, and fruit; takes sunflower seeds, suet, and peanuts at feeder.
HOUSE/BEWICK'S WREN	Insects and spiders. Will take finely ground suet and nuts from ground.
EASTERN/WESTERN BLUEBIRD	Insects and fruit; at feeder likes peanut butter, cornmeal, and mealworms.
ROBIN	Worms, insects, and fruit; cakes, bread, raisins, and apples at feeder. Likes most berry bushes.
NORTHERN MOCKINGBIRD	Insects and fruit; takes suet and raisins at feeder. Likes fruits of elderberry, blackberry, and red cedar.
BROWN THRASHER	Insects and berries, sometimes nuts and corn; will take suet, wheat, millet, and sunflower seeds from feeder.
CEDAR WAXWING	Fruit and insects; at feeder will take raisins, currants, and apples. Particularly fond of berries from mountain ash, pyracantha, privet, cedar, and mulberries.
NORTHERN CARDINAL	Insects, fruit, and seeds, especially of pine; comes to feeder for cracked wheat and sunflower seeds.
SONG SPARROW	Insects, seeds, and fruit; comes for seed mixtures placed on ground.
WHITE-CROWNED SPARROW	Seeds and insects. Comes to feeder for millet and seed mixtures.
DARK-EYED JUNCO	Insects and seeds; takes seed mixtures at feeder, also attracted by seeds of zinnias and cosmos.
HOUSE FINCH	Seeds, fruit, and insects; takes mixed seeds, fruit, and scraps from feeders.
PURPLE FINCH	Seeds, insects, and fruit; visits feeder for hemp, millet, and sunflower seeds.
BALTIMORE ORIOLE	Insects and fruit. Comes to feeder for apple, orange, banana, grapes, and suet.
SCARLET TANAGER	Insects and berries; comes to feeder for peanut butter, cornmeal, apple, banana, cherries, and raisins.

Pecking order

As well as simply looking at or listening to the birds in your backyard, it can be interesting to study their behavior, especially when feeding. Many exhibit a pecking order either within their own species or between different species. Among birds of the same species, males will often dominate females and not let them feed at the same time, while young birds may be driven off by older ones. There are always some species that appear more aggressive than others, displacing the more timid ones from the feeders. Occasionally a bird may make a feeder, or even a berry bush, its "own" — robins sometimes do this — and will drive off all intruders. If you notice that some birds are monopolizing a feeding station, try to spread out the food so that others can get to it. Don't just put seed in feeders or on a platform — throw some on the ground, as well, broadcasting it rather than placing it in one small area.

AN ASSORTMENT OF NEST BOXES

While a garden can provide relatively natural nest sites for birds that build in bushes, trees, and grass, those birds who nest in tree holes rarely find a natural site. This is where bird boxes are indispensable. Boxes can be made to suit a wide range of species, and should always be positioned in

Any flat, protected structure is likely to attract the American robin to nest in your backyard.

appropriate locations. When constructing a box, it is important to get the hole size correct for the species you want to attract. The box should have the right internal dimensions, as well — not small. Lastly, the hole should not be too close to the floor nor too high up, although some birds may fill the box with nesting material until a suitable height is reached. Once built, the box must be positioned correctly as some birds will nest higher than others. Most boxes are simply fixed securely to a tree trunk, though purple martins like their boxes on a post. Bluebirds, wrens,

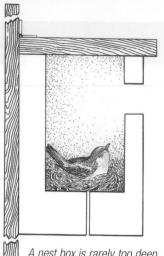

A nest box is rarely too deep, as the bird will fill the space with nesting material to its preferred height. Too shallow a box might allow a predator to reach in and take the young.

A nest box should never be placed facing into the prevailing wind. If the box is tilted slightly downwards, this will reduce the risk of rain getting in and chilling the nestlings.

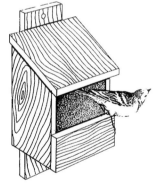

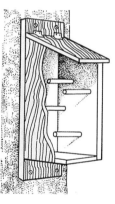

A hole-fronted box is good for house finches, chickadees, wrens, and titmice.

An open-fronted box will appeal to some flycatchers. It should be partly hidden in foliage to keep away predators.

Some birds will use nest boxes as winter roosts. A roosting box can be made with an entrance hole and perch.

Lengths of doweling are used for perches inside the roosting box so that many birds can enter.

Artificial mud nests for cliff swallows can be placed under the eaves. Weighted strings will stop house sparrows from moving in.

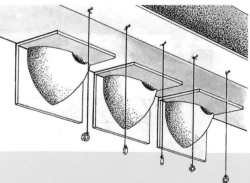

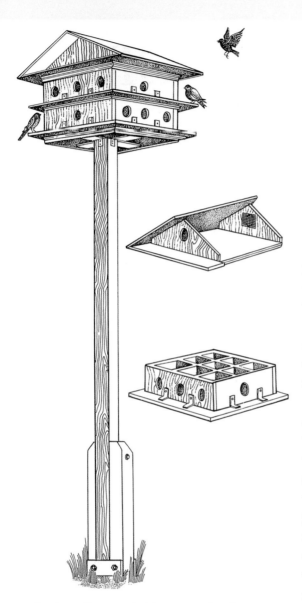

Purple martins are colonial nesters and prefer an apartment block to a single nestbox. Their high-rise accommodation is about 20 ft from the ground and must be lowered for cleaning each year.

chickadees, titmice, nuthatches, and swallows will all use boxes, while barn swallows, phoebes, and robins prefer shelf nests (simply boxes with no fronts and sometimes no sides). Remember that birds are territorial when breeding, and if you fill a garden with dozens of boxes, it is likely that only a few will be used.

Boxes should be cleaned out and repaired at the end of the nesting season. Removing an old nest will often remove parasites and their eggs, which are lying dormant until the next breeding season, waiting to infest the young. If nest boxes are left up outside the breeding season, some birds such as wrens, titmice, chickadees, nuthatches, and bluebirds may roost in them.

Gardens that attract songbirds will always be full of activity throughout the year. Pleasure can be derived from watching the birds feed in winter, seeing them courting or squabbling over territories in spring, and raising their young in summer. The arrival of the first migrants in spring and fall can be most exciting, and the occasional rare visitor will always be rewarding. A bird garden need not be untidy and overgrown, and will contain more life than any carefully manicured, weed- and pest-free example of the horticulturalist's art.

Brown creepers normally nest in a narrow tree crevice. A strip of bark fixed to a tree may be used for nesting or roosting.

A more formal creeper box can be made from wood with a small hole at the side close to the bark of the tree.

A number of species utilize birdboxes for nesting. Tree swallows often get help from the young of the previous year.

All sorts of different nest sites may be used by wrens, such as a coconut with a hole in it or a hanging flowerpot.

A more open nesting tray with a roof may be used by American robins and phoebes if placed on the side of a house.

Birds need material to build and line their nests. A basket full of string, wool, hair, and feathers will help them.

NESTING REQUIREMENTS

Measurements of nest boxes are given as width x depth x height; nest hole dimensions refer to the diameter of the nest hole.

SPECIES	NESTING/NEST BOX
GREAT-CRESTED FLYCATHER	Nests naturally in tree holes. Box 6 x 6 x 10 in with 2 in nest hole, placed above 6 ft.
TREE SWALLOW	Uses natural tree holes. Box 5 x 5 x 6 in with 1½ in hole, placed high on tree or under eaves of old building.
BLUE JAY	Nest of twigs placed 10 to 15 ft high in trees, bushes, vines. Oak and beech trees provide both food and nest sites.
BLACK-CAPPED CHICKADEE	Usually excavates hole in rotten branch. Box 4 x 4 x 8 in with 1⅛ in hole, placed above 8 ft.
TUFTED/OAK TITMOUSE	Natural cavities. Box 4 x 4 x 8 in with 1¼ in hole placed above 8 ft.
HOUSE/BEWICK'S WREN	Nests in almost any cavity. Box 4 x 4 x 8 in with 1 in hole at 5 to 10 ft.
EASTERN/WESTERN BLUEBIRD	Natural tree holes. Box 5 x 5 x 8 in with 1½ in hole placed at 5 to 10 ft.
ROBIN	Nests on tree forks, in bushes, and will use ledges and shelves.
NORTHERN MOCKINGBIRD	Nests in bush, vine or tree and tangles.
BROWN THRASHER	Nests in low bushes such as white currant, lilac, privet, plum, and in brush piles.
CEDAR WAXWING	Nests in trees and shrubs where food is plentiful.
NORTHERN CARDINAL	Nests in trees, thickets, and vines. In garden likes young evergreens, rosebushes, and and honeysuckle.
SONG SPARROW	Nests in long grass and low bushes.
WHITE-CROWNED SPARROW	Nests on ground, in grass or under shrubs, sometimes in small conifer.
DARK-EYED JUNCO	Nests on ground under tree roots, in brush piles and under house gables.
HOUSE FINCH	Nests in holes in trees and buildings. Box 6 x 6 x 6 in with 2 in hole placed at 8 to 12 ft.
PURPLE FINCH	Nests high in dense conifers and deciduous trees.
BALTIMORE ORIOLE	Suspends nest from branch of maple, elm, poplar, or conifer.
SCARLET TANAGER	Nests on branch of large oak, ash, or maple.

The Directory
of Songbirds

At-a-glance identifier

Tyrant flycatchers:
Tyrannidae
Active birds that snap up insects by darting out from a perch; songs are varied, often harsh or shrill.

Eastern phoebe 54

Olive-sided flycatcher 60

Eastern kingbird 50

Say's phoebe 56

Vermillion flycatcher 62

Eastern wood pewee 52

Great-crested flycatcher 58

Flycatchers:
Empidonax
Look-alike small flycatchers; songs have different two-syllabled notes.

Willow flycatcher 64

Larks:
Alaudidae
Ground-nesting birds with musical songs, often given in flight.

Horned lark 66

Swallows:
Hirundinidae
Fast-flying birds with a song of rapidly delivered chirrups or twitters.

Purple martin 68

Tree swallow 70

Jays and crows:
Corvidae
Bold medium- to large-sized birds with stout bills. Jays utter both harsh and soft notes, crows tend to deliver raucous caws.

Gray jay 72

Blue jay 74

American crow 76

Titmice and chickadees:
Paridae
Small acrobatic birds; songs made up of clear repeated notes.

Black-capped chickadee 78

Boreal chickadee 80

Tufted titmouse 82

White-breasted
nuthatch 88

Wrens:
Troglodytidae
Tiny, rounded brownish
birds with loud beautiful
songs and harsh alarm calls.

Oak titmouse 84

Canyon wren 92

Nuthatches:
Sittidae
Small tree-climbing birds
whose songs tend to be
nasal-toned whistles.

Creepers:
Certhiidae
Small, camouflaged tree-
climbers, songs made up of
thin, hissing notes.

Cactus wren 94

Red-breasted
nuthatch 86

Brown creeper 90

Bewick's wren 96

Dippers:
Cinclidae
Similar to large wrens,
associated with mountain
streams; rich bubbling songs
and sharp calls.

Blue-gray
gnatcatcher 108

House wren 98

Dipper 104

Winter wren 100

Kinglets and gnatcatchers:
Sylviidae
Tiny, active birds with
high, thin calls.

Bluebirds and thrushes:
Turdidae
Bluebirds snap up insects
in the air, thrushes tend
to forage on the ground;
both species have rich,
varied songs.

Ruby-crowned
kinglet 106

Eastern
bluebird 110

Marsh wren 102

Mountain
bluebird 112

Wood thrush 120

Mimic thrushes:
Mimidae
Medium-sized
birds with distinctive,
often loud, songs.

Veery 114

Robin 122

Mockingbird 128

Swainson's
thrush 116

Varied thrush 124

Gray catbird 130

Hermit thrush 118

Townsend's
solitaire 126

Brown
thrasher 132

Wrentits:
Chamaeidae
Sparrow-sized birds with heads like titmice and the long tail of wrens; the song is rapidly delivered.

Wrentit 134

Pipits:
Motacillidae
Slender, lark-like birds with high, airy flight songs.

Water pipit 136

Waxwings:
Bombycillidae
Slim, crested birds with thin, high songs.

Cedar waxwing 138

Shrikes:
Laniidae
Medium-sized birds with a stout, hooked beak. The melodious songs are interspersed with harsh notes.

Loggerhead shrike 140

Vireos:
Vireonidae
Arboreal birds with slow deliberate movements; their distinctive songs usually have repeated phrases.

Warbling vireo 142

Red-eyed vireo 144

Bell's vireo 146

Wood warblers:
Parulidae
Small, active, restless
birds; most songs are varied,
often high-pitched buzzes
and trills.

Yellow warbler 152

Pine warbler 158

Kentucky
warbler 148

Chestnut-sided
warbler 154

Blackpoll
warbler 160

Northern parula
warbler 150

Black-throated
warbler 156

Black-and-white
warbler 162

Wilson's warbler
164

Ovenbird 170

Common
yellowthroat 176

American
redstart 166

Louisiana
waterthrush 172

Yellow-breasted
chat 178

Prothonotary
warbler 168

Connecticut
warbler 174

Tanagers:
Thraupidae
Birds of tropical origin;
males are brilliantly colored
and sing hoarse robinlike
songs from treetops; also
have dry, staccato calls.

Grosbeaks, finches, buntings:
Fringillidae
Small birds with short,
conical bills. Most are
arboreal, sparrows (also
included in the group) are
more terrestrial. Grosbeaks
warble; finches tend to chirp;
bunting songs are often high
and lively; and sparrows
have musical whistles.

Rose-breasted
grosbeak 190

Summer
tanager 180

Western
tanager 182

Pyrrhuloxia 186

Black-headed
grosbeak 192

Scarlet tanager 184

Northern
cardinal 188

Lazuli bunting 194

Indigo bunting 196

Eastern towhee 202

Chipping
sparrow 208

Dickcissel 198

Bachman's
sparrow 204

Lark bunting 210

Green-tailed
towhee 200

Tree sparrow 206

Henslow's
sparrow 212

Fox sparrow 214

White-throated
sparrow 220

House finch 226

Vesper sparrow 216

White-crowned
sparrow 222

American
goldfinch 228

Song sparrow 218

Dark-eyed junco 224

Red crossbill 230

Pine siskin 232

Red-winged
blackbird 238

Eastern
meadowlark 244

Purple finch 234

Yellow-headed
blackbird 240

Baltimore oriole 246

Blackbirds and orioles:
Icteridae
Medium-sized birds with
strong, pointed bills.
Blackbirds have loud,
coarse songs; orioles use
piping whistles.

Rusty blackbird 236

Bobolink 242

Orchard oriole 248

How to use the directory

The great range of notes and rhythms on display in birdsong can be confusing, but once mastered, the language adds a whole new dimension to bird study. This directory helps to do this, featuring the songs and calls of 100 songbirds.

Information on when and where all these birds sing is accompanied by interesting aspects of feeding and breeding behavior. With the birds grouped in families, it is easy to make out the broad outline of any common song features for each group — for example, the sweet, clear notes of the thrushes, the buzzy, wheezy notes of sparrows, and the thin high-pitched songs of many of the warblers.

Scientific name of species (in Latin)

Common name of species (in English)

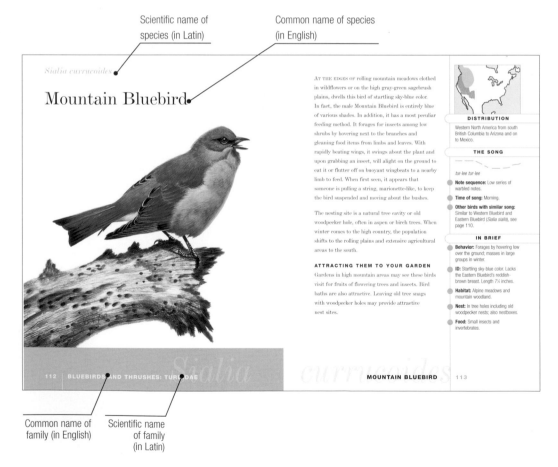

Sialia currucoides

Mountain Bluebird

AT THE EDGES OF rolling mountain meadows clothed in wildflowers or on the high gray-green sagebrush plains, dwells this bird of startling sky-blue color. In fact, the male Mountain Bluebird is entirely blue of various shades. In addition, it has a most peculiar feeding method. It forages for insects among low shrubs by hovering next to the branches and gleaning food items from limbs and leaves. With rapidly beating wings, it swings about the plant and upon grabbing an insect, will alight on the ground to eat it or flutter off on buoyant wingbeats to a nearby limb to feed. When first seen, it appears that someone is pulling a string, marionette-like, to keep the bird suspended and moving about the bushes.

The nesting site is a natural tree cavity or old woodpecker hole, often in aspen or birch trees. When winter comes to the high country, the population shifts to the rolling plains and extensive agricultural areas to the south.

ATTRACTING THEM TO YOUR GARDEN
Gardens in high mountain areas may see these birds visit for fruits of flowering trees and insects. Bird baths are also attractive. Leaving old tree snags with woodpecker holes may provide attractive nest sites.

DISTRIBUTION
Western North America from south British Columbia to Arizona and on to Mexico.

THE SONG

tur-lee tur-lee

Note sequence: Low series of warbled notes.
Time of song: Morning.
Other birds with similar song: Similar to Western Bluebird and Eastern Bluebird (*Sialia sialis*), see page 110.

IN BRIEF
Behavior: Forages by hovering low over the ground; masses in large groups in winter.
ID: Startling sky-blue color. Lacks the Eastern Bluebird's reddish-brown breast. Length 7¼ inches.
Habitat: Alpine meadows and mountain woodland.
Nest: In tree holes including old woodpecker nests; also nestboxes.
Food: Small insects and invertebrates.

112   BLUEBIRDS AND THRUSHES: TURDIDAE

Sialia   currucoides

MOUNTAIN BLUEBIRD   113

Common name of family (in English)

Scientific name of family (in Latin)

THE SONG

The rhythmic flow of a bird's song is given for each species using simple linear notation. The length of the pause between notes depicts the rapidity of the song.

The base components are as follows:

Rising pitch	
Dropping pitch	
Long note	
Fast rising inflection and sudden drop	
A rollercoasting effect	
Very rapid	
Long notes and long pauses	
Short notes and long pauses	

THE MAPS

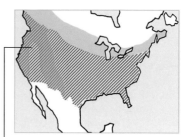

Outlined maps of North America give distribution information. The maps show summer distribution (in green), winter distribution (in diagonal hatching) and all year distribution (in green with diagonal hatching).

Summer distribution	
Winter distribution	
Year-round distribution	

Carduelis pinus

Pine Siskin

AS THE LEAVES OF FALL appear throughout the East, the crisp air carries the high-pitched "*sreeee – sreeee – ji–ji–jit – sreee*" of the first Pine Siskins. Often they will be seen in small groups bobbing along in the characteristic flight of the goldfinch family. They rise and fall in this undulating pattern, then quickly swing around and plummet downward, landing in the treetops.

The slim, dark bill separates them from the other groups of heavier billed finches. They use it to manipulate the catkins in trees until a shower of shucked parts filters from the tree. Then, as if a silent signal had been given, they explode from the treetops and are gone from sight.

The number of siskins reaching the southern edge of their range varies greatly. During some winters, thousands will be seen. Such invasion years often lead to residual populations that at times nest far south of their normal nesting range. In the western portion of North America, the mountainous areas and high plateaus host the species year-round.

ATTRACTING THEM TO YOUR GARDEN

Their favorite food is thistle seed and a thistle seed feeder will be emptied rapidly as a progression of birds occupy the pegs throughout the day. They are also attracted to trees with catkin-type fruit such as an alder or birch. Sunflower seeds are also taken.

DISTRIBUTION
Across Canada, northern U.S. and in the western mountains.

THE SONG

swee swee see jit jit jit see see

Note sequence: Very reedy in quality and always incorporating the swee notes typical of the family.

Time of song: All day.

Other birds with similar song: Close to American Goldfinch (*Carduelis tristis*, see page 228, but more raspy.

IN BRIEF

Behavior: Gregarious, relatively tame, often forms large flocks with other small finches.

ID: Trim with dark brown streaking. Distinct yellowish wing bars with yellow also on sides of tail. Length 5 inches.

Habitat: Coniferous woodlands and adjacent mixed woodland. In winter, backyards, gardens, parks.

Nest: A shallow cup of grasses, rootlets, and mosses lined with feathers and fur.

Food: Seeds and some insects. Thistle seeds preferred at feeder.

232 | GROSBEAKS, FINCHES, BUNTINGS: FRINGILLIDAE

Carduelis *pinus*

PINE SISKIN 233

Eastern Kingbird

Tyrannus

Northern New England west to northern British Columbia then south across the plains to eastern Texas and Florida.

THE SONG

kazeeh kahzeeh-kip-kip dzypper dzypper

- **Note sequence:** A loud series of twitters and harsh notes.

- **Time of song:** All day.

- **Other birds with similar song:** All Flycatchers have twittering notes, but none have a true sweet song.

IN BRIEF

- **Behavior:** Flies out from treetops on quivering wings; usually solitary; very defensive of nesting territory.

- **ID:** Gray above, dark head, black tail, white underparts, and narrow white wing bars. A small orange patch on the crown is partly concealed and difficult to see in the field. Length 8 inches.

- **Habitat:** Open areas such as field edges, forest clearings, farmlands, orchards, gardens; often near water.

- **Nest:** Large loose cup of grasses, weeds, and other plant material, lined with plant down.

- **Food:** Winged insects, especially bees, occasionally berries.

FLYCATCHERS IN GENERAL pose a problem in identification for the beginning birder. Therefore, it is welcome to find one that is easy to identify by sight, sound, and action. The white band on the end of the tail is found on no other North American flycatcher.

The bird's display flights contrast with the more directed flight of a bird darting off a perch to capture a flying insect. With stiffened wingbeats, it makes this quivering flight calling "*zeek-zeek-zeek*" followed with "*kip-kip-kip-dzypper-dzypper.*" When completing the display flight, or when incensed by the presence of a predator, the crown feathers are raised and the fiery rust-red crown feathers can be seen. It is this crown that gives the group its name of "kingbirds."

The nest is placed well off the ground in the crotch of a tree. In defense of this site the Kingbird dive-bombs ground predators.

ATTRACTING THEM TO YOUR GARDEN

The Eastern Kingbird's favorite perch is atop a tree in an orchard, garden, open field edge, or along a waterway. Water availability is important, and nest building materials such as short pieces of yarn and string often attract them. Favored nesting trees are often fruiting trees, which attract a host of insects used for food.

tyrannus

EASTERN KINGBIRD

Contopus virens

Eastern Wood Pewee

IN THE DRY UPLAND woods of early summer, the long drawn-out *"peeeee-weee"* is a familiar sound. Tracking the singer down will lead to a small, drab brown bird sitting in a vertical position, often on the dead stub of a branch fairly high off the ground. The bird uses the perch as a command post for spotting insects and quickly darts out to grab them before returning to its favorite spot.

When not in song, the bird may be difficult to see due to its sedentary ways. The nest is even more difficult to locate. It is placed on the upper surface of the limb of a tree and usually at a bend, and looks like a small knot on the limb.

The species is interesting in that it varies its song depending on the time of day. Most of the day it is a repetitious *"pee-a-wee"* which at times is followed by a slurred *"pee-oooo."* However, the early morning and low light levels of late afternoon are marked by an explosive part to the song, with an additional A sharp, rising *"a-deedit."*

ATTRACTING THEM TO YOUR GARDEN

Unless you live in a woodland setting where large trees are favored as nesting sites, this species may appear in migration only. Flowering trees, which host a wide selection of visiting flowers, often attract.

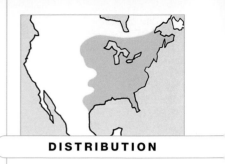

DISTRIBUTION

Eastern half of the U.S. to northern Florida.

THE SONG

peee-a-weeeee pee-a-weeeee

Note sequence: A loud, long drawn out and often mournful sound with a rising inflection at the end.

Time of song: All day throughout summer.

Other birds with similar song: Some similarity to the breeding call of the Black-capped Chickadee (*Poecile atricapillus*), see page 78.

IN BRIEF

Behavior: Sedentary, calls from broken branch stub; frequents the area just below the tree canopy.

ID: Brown with two wing bars and orange-yellow lower mandible. Length 6¼ inches.

Habitat: Woodlands, orchards.

Nest: Shallow well-disguised cup of mosses and lichens, looks like a knot of a limb.

Food: Insects, which are snapped up as it sallies forth from a dead limb perch.

virens

EASTERN WOOD PEWEE

Sayornis phoebe

Eastern Phoebe

When the northeastern woodlands are still drab gray but hints of buds tint the branches reddish, the emphatic call of *"fee-beep – fee-beep"* marks the arrival of the first flycatcher of the spring, the Eastern Phoebe. When found, the songster is usually in a bolt upright position, tail bobbing up and down, and rapidly looking about for insects.

This species has adapted well to the structures of man. Although they can still be seen nesting within the mixed upland woods on cliff overhangs and rocky outcrops, they also fit right into a suburban situation. Understreet culverts are favorite sites, as is any bridge near a suitable woodland habitat. Buildings are used and overhangs of barns and garages are also favored. If the nest is made in a well-protected site, it will be upgraded and used for many years in succession.

In the fall, young birds show a marked lemon tint to the underparts, and indistinct buff-edged wing bars. No other flycatcher at this time shows these features. They linger well into the fall and some even attempt overwintering, shifting to a berry diet.

ATTRACTING THEM TO YOUR GARDEN

Flowering shrubs and trees will attract insects and hence food for this species. A nesting shelf of a flat surface with sheltering roof attached to a building at a distance well-off will appeal.

DISTRIBUTION

Eastern portion of North America from the Great Plains to north New England and south to eastern Texas and northern Florida.

THE SONG

fee-beep fee-beep fee-beep

- **Note sequence:** Emphatic and two-parted with a sharp inflection upward on the second part.

- **Time of song:** Morning.

- **Other birds with similar song:** In spring, the Black-capped Chickadee gives a loud "feee-bee" whistle, but not the sharp "phoebee" of this Flycatcher.

IN BRIEF

- **Behavior:** Cocks tail up and down while sitting; attracted to water.

- **ID:** Gray above and pale below, the head shows very dark in contrast to the back. Lack of eye ring and wing bars. Length 7 inches.

- **Habitat:** Woodlands, parks, gardens, backyards; common around buildings.

- **Nest:** A mud and moss cup under eaves, bridges, culverts or rock faces.

- **Food:** Mainly insects and spiders with berries in winter.

phoebe

Sayornis saya

Say's Phoebe

Sayornis

THE SAY'S PHOEBE'S RANGE extends from the edge of the Arctic Circle south to the Mexican border in Arizona and down well into Mexico. Throughout, the bird prefers the dry, open areas, canyons, and cliffs that form a network over this vast range.

Within these canyons the explosive *"pee-yeet"* call is magnified and echoed, making the exact location of the singing bird difficult to spot. When sighted, the bird is found perched atop a boulder or scrub-bush pumping its tail up and down and often fanning it open with rapid flicks.

Near the nest site the *"pi–pit pit–pit seeer"* song, after the explosive call notes, is very often delivered as the bird hovers in the air before returning to its roost spot. In underdeveloped areas, cliff faces and rocky outcrops are the selected sites to place the nest. In developed areas, a flat site is preferred, under the overhang of a roof, under a bridge or, in the Southwest, even in old mine shafts. This highly migratory species overwinters in Mexico and further south, and stray birds, mainly immatures, do from time to time show up on the Eastern seaboard.

ATTRACTING THEM TO YOUR GARDEN

A bird of transition, they do come into gardens in the winter, especially in southern parts of their range where they often are permanent. They will take small fruits of shrubs and trees in winter and enjoy sites where water is available.

DISTRIBUTION

A wide range from central Alaska to the edge of the plains and south to the Arizona/west Texas border

THE SONG

_ / _ / _ /

pee-yeet pee-yeet

- **Note sequence:** Explosive and upward pitched.
- **Time of song:** Mainly morning and near dusk.
- **Other birds with similar song:** Similar in sequence to Vermillion Flycatcher (*Pyrocephalus rubinus*), see page 62, but much louder.

IN BRIEF

- **Behavior:** Active and restless; flutters to ground to feed; bobs tail as it sits.
- **ID:** Gray with darker gray on top of head. Black tail and rust-colored belly and undertail feathers. 7–8 inches long.
- **Habitat:** Open and rocky country.
- **Nest:** Cup of mud balls and grasses lined with feathers, placed under an overhang of rock or habitation where shelf is available. Will use a nesting platform placed on the wall of a building.
- **Food:** Insects, and will glean spiders from rock surfaces, berries taken in winter.

saya

Great-crested Flycatcher

Myiarchus

THIS SPECIES IS A COMMON RESIDENT of open woodlands in the eastern half of the United States. However, because it prefers the upper areas of oaks, hickories, and maples, it is not always that easy to see. Indeed, the distinctive *"weeerrrup"* call is often the only give-away of this bird's presence. This large, distinctively marked flycatcher is usually sitting on a dead stub, within the canopy, in an upright position, and constantly moving its head around in search of insect prey. This is one of the few flycatchers that uses a natural cavity for nesting. Old woodpecker holes are also used and the birds will also take readily to nesting boxes placed in their territory.

With a lack of winter food supply in the north, the population migrates south to winter.

ATTRACTING THEM TO YOUR GARDEN

If you live near a woodland edge, bird houses of proper size and placed well off the ground may attract this species to nest. The species has an odd habit of placing shed snake skins in its nest. If you should find any, place them over a tree limb as an enticement for nesting. Flowering trees attract insects as a food source.

DISTRIBUTION

Eastern half of the U.S. from the Dakotas to New England south to Texas and Florida.

THE SONG

weeerrrrup weeerrrrup treert treert

- **Note sequence:** A rising inflection weerrup or a loud echoing treert, sounding similar to a whistle being blown.

- **Time of song:** Morning.

- **Other birds with similar song:** Most of the Myiarchus Flycatchers have rolling, rough calls.

IN BRIEF

- **Behavior:** Frequents treetop canopy; usually sedentary; very aggressive.

- **ID:** Bushy crest and large broad bill. Brown upper parts, dark gray throat down to mid-chest with lemon-yellow underparts. Tail and wing are marked with cinnamon, best seen when spread. Length 8½ inches.

- **Habitat:** Woods and orchards.

- **Nest:** Hole nester, nest composed of sticks, grasses, leaves, and other plant material. Often snakeskin or paper in the lining.

- **Food:** Insects.

Contopus borealis

Olive-sided Flycatcher

Contopus

THE SONG OF THE OLIVE-SIDED FLYCATCHER, once heard, is not forgotten. A very clear and distinctive "*quick-three-beers*" is often followed by a series of "*pit-pit-pit*" call notes. The bird's favored breeding haunts are the cool boglands and evergreen forests of the north. The best way to locate the songster is to scan the tall dead spires of spruce where this large flycatcher may be seen perched on the very top in an upright posture. It may dart out into the air to grab insects but will quickly return to the spire to call and sing. Dusty gray sides wrap in toward the central breast giving the bird the appearance of wearing a "vest."

With the onset of fall, the birds head south on a rather long journey that takes them many hundreds of miles away to their wintering grounds, the volcanic hillsides and cool ravines of northern South America.

ATTRACTING THEM TO YOUR GARDEN

Only seen as a transient in most yards, one possible attractant is to leave dead snags standing to act as feeding perches as they pass through.

DISTRIBUTION

Across Canada and south Alaska and in the mountains to California, Arizona and northern New England.

THE SONG

quick-three-beers

● **Note sequence:** A distinct three-parted series; the phrase is drawn out.

● **Time of song:** All day.

● **Other birds with similar song:** The whip, whip note sounds similar to the Great-crested Flycatcher (*Myiarchus crinitus*), see page 58.

IN BRIEF

● **Behavior:** Sings from an exposed perch; strongly defends nest site.

● **ID:** Large and stocky with short tail. Dull white breast runs into white throat. Smudgy gray sides of breast. Olive-green back, white underparts and dark green, streaked flanks. White tufts on either side of rump show well in flight. Length 7½ inches.

● **Habitat:** Dead trees in coniferous forest, especially spruce in north of range and pitch pine in the south.

● **Nest:** Shallow cup of twigs covered with mosses or lichens, placed well out on slender limb of conifer.

● **Food:** Insects.

Vermillion Flycatcher

THE MALE VERMILLION FLYCATCHER is the showpiece, glowing vermillion with black back, wings, and tail and a jet-black mask, although the female, with her mouse-brown back, striped underparts, and blush rouge sides and underbelly is also attractive. During the breeding season, the male will often do a flutter flight when the female is near and then the soft, musical song of *"hit-a-see hit-a-see"* can be heard.

When found, they are usually fairly approachable. Sitting atop a dead snag, it has a habit of constantly pumping and fanning its tail, head jerking about as it scans for insects.

Though found in a variety of habitats, they prefer areas in the generally arid Southwest where there is constant water. The nest is tucked tightly into the fork of a tree crotch, often along waterways.

ATTRACTING THEM TO YOUR GARDEN

This flycatcher lives in thickets near water in the Southwest. Stream beds, river edges, water-towers, and drinking troughs are all attractive to them. Be sure water is available, as well as fruiting trees and shrubs where visiting insects can form a food base. Have been known to nest in boxes.

Southern portion of Arizona, New Mexico and Texas; strays to California and east to Florida in winter.

THE SONG

hit-a-see hit-a-see

● **Note sequence:** A series of rather explosive notes in sequence with rising inflection on the last portion of the three-parted song.

● **Time of song:** Morning and evening.

● **Other birds with similar song:** Similar to song of Say's Phoebe (*Sayornis saya*), see page 56.

IN BRIEF

● **Behavior:** Generally a high canopy or thick scrub feeder but will dart out to feed or drop to the ground.

● **ID:** The male has a bright red crown and underparts. The adult female has gray-brown plumage, a finely streaked white throat and breast, and a pink belly and undertail. Length 6 inches.

● **Habitat:** The scrub and trees that fringe watercourses in arid areas.

● **Nest:** Lichen-covered cup in crotch of branch.

● **Food:** Insects.

Willow
Flycatcher

Empidonax

THE WILLOW FLYCATCHER'S SONG is most diagnostic: a wheezy *"fitz-bew"* with the emphasis on the first part. It lacks the well-defined eye-ring of its look-alike relatives, has a white throat that contrasts to the olive-tinted breast, and the underbelly shows a yellowish cast.

The species frequents a wide variety of habitat over its extensive range: pastureland, brushy clearings, stream edges, orchards, and on up to rolling mountain meadows. In general it is found in a drier situation than its closest look-alike, the Alder Flycatcher. In fact, for many years the two species were considered to be one.

In migration, these birds are likely to appear in almost any habitat. They sit without moving for long periods and are often overlooked. During this time, listen for a liquid *"whit-whit"* call and you could be well-rewarded.

ATTRACTING THEM TO YOUR GARDEN

As its name suggests, this flycatcher prefers willow, as well as alder thickets. They are seen in yards as a transient, like all flycatchers. Any plantings that attract insects offer a food source as they pass through.

DISTRIBUTION

New England to south Appalachians, west to northwest to southwest. Avoids the Great Plains.

THE SONG

fitz-bew fitz-bew fitz-bew

Note sequence: Snappy and two-parted, it ends with a wispy quality.

Time of song: All day.

Other birds with similar song: Other Empidonax Flycatchers have wheezy songs but their diagnostic phrasing separates the species.

IN BRIEF

Behavior: Sedentary, darts out to capture prey; snaps head back when it calls.

ID: Pale eye-ring and dark olive-brown head. White throat, olive-tinted breast, and yellowish underbelly. Length 5½ inches.

Habitat: Wet meadows and field edges.

Nest: Cup of grasses and rootlets covered with silken materials fixed by spider webbing. Suspended in the fork of a branch.

Food: Insects.

Eremophila alpestris

Horned Lark

THE HORNED LARK is a common bird throughout North America, except in the very limits of the Southeast. It nests from far above the Arctic Circle on down through Central America. The base pattern is always present — tawny brown with distinct mask through the eyes, a black bib, and two horn-like feather groups in breeding plumage. Very often, their presence is announced by their sweet tinkling song given from high overhead as they circle about. They scuttle about very low to the ground in search of seeds, grass, and sedge seeds in particular, and are at home both on a wind-blown rocky ridge of the Alaskan Range and a freshly turned-over farm field in Texas. The young birds are black and white in color, often giving a zebra pattern in appearance.

Overwintering flocks moving south to the U.S. from their tundra breeding grounds usually head for wind-swept beaches or farm fields. In such places as southern California, it is not uncommon to see fields with thousands of these birds. These disperse by late winter, and in early spring the tinkling song is once again heard over a significant portion of the northern hemisphere.

ATTRACTING THEM TO YOUR GARDEN

If you live near the shore or open fields, this species might drop in if your area is large enough. Large plowed areas are often attractive. Confined areas will not attract them.

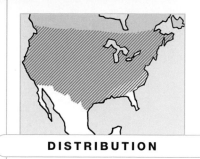

DISTRIBUTION

Widespread throughout North America from Alaska to the Mexican border.

THE SONG

peet tweedle ee dee deet

- **Note sequence:** A soft tinkling series of notes usually delivered from high overhead.

- **Time of song:** All day.

- **Other birds with similar song:** The tinkling notes of Tree Swallows (*Tachycineta bicolor*), see page 70, might be confused with this species.

IN BRIEF

- **Behavior:** Sings in the air and on the ground; shuffles about when feeding; gregarious.

- **ID:** Horned larks vary in color from pale gray to reddish-brown, but all have the distinctive black breast mark and facial pattern. Many have a yellow face. Length 7½ inches.

- **Habitat:** Tundra, alpine areas, shorelines, agricultural land.

- **Nest:** Cup of grasses on the ground, at the base of a rock or clump of grass.

- **Food:** Seeds, insects, and other small invertebrates.

alpestris

Progne subis

Purple Martin

Progne

No NORTH AMERICAN SPECIES relies so much on man for housing as does the Purple Martin. Prior to the human development of the continent, the birds nested in natural tree cavities and rock crevices, which remain a last resort in some areas, and colonies do inhabit the giant trees of western forests. But in most regions, man-made structures are the only sites occupied — some accommodating over 1,000 pairs of birds.

The Purple Martin has a rich, loud, liquid song and is graceful in flight, executing spectacular aerial maneuvers. As an exclusive insect eater, it moves south for the winter, not just a short way but on into South America where its complete winter area has as yet to be fully described. They return north early, a few birds trickling in to signal the arrival of the main influx within a couple of weeks. The nests are checked and first investigations made. By early April, the skies are again alive with birds.

ATTRACTING THEM TO YOUR GARDEN

The Purple Martin eats insects, especially mosquitoes, which it catches on the wing. It can be attracted to specially built communal nesting boxes and special grounds. Be sure water is nearby and a large open swoop area is present around the nest boxes. Take down boxes in winter so sparrows do not take them over.

DISTRIBUTION

The eastern half of North America with scattered populations in the Pacific northwest, California, and Arizona.

THE SONG

gurgling bubbley

- **Note sequence:** Very liquid in quality with a downslurring of notes.

- **Time of song:** All day.

- **Other birds with similar song:** Guttural songs of other swallows might be confused with this species.

IN BRIEF

- **Behavior:** Flies with rapid wingbeats alternating with sailing flight, often similar to a great arc, swoops low over open areas; strongly defends territory.

- **ID:** Males sport all-purple iridescent plumage. Length 5 inches.

- **Habitat:** Open woodlands, farmlands near lake edges, saguaro deserts.

- **Nest:** In old woodpecker holes, natural tree cavities, and man-made nesting boxes.

- **Food:** Flying insects, especially mosquitoes.

subis

Tachycineta bicolor

Tree Swallow

Tachycineta

Swirling masses of thousands of these birds are not an uncommon sight in the fall as they head south on migration, bound for the extreme southern portions of the United States, Mexico, and Central America. En route, fruiting bushes such as bayberry bend to the ground under the weight, as every fruit is picked clean.

This bird is a very successful nester and natural cavities and birdhouses are taken readily. The birds favor open fields and meadows; flooded swamplands with dead trees riddled with old woodpecker holes are other favored nesting sites. The loud repeated twitterings at these sites is an early spring ritual.

The green of the bird's iridescent back turns to a black mask under the eye and onto the cheek. This allows rapid separation from the look-alike Violet-green Swallow, which has white wrapping around to the back of the eye.

When gathering material for the nest cavity, the swallows' mastery of the air can be truly appreciated as they drop and swoop, gathering materials.

ATTRACTING THEM TO YOUR GARDEN

Provide several nesting boxes to allow choice. Put out feathers to be used as nesting material. Plantings such as bayberry and red cedar will provide food in fall migration.

DISTRIBUTION

Across North America from Alaska to Newfoundland south to the Carolinas and central Arizona.

THE SONG

tslee tslee tslee tslee

- **Note sequence:** A series of rapid, lispy notes.
- **Time of song:** All day.
- **Other birds with similar song:** All the swallows have sweet to raspy short burst twitterings.

IN BRIEF

- **Behavior:** Skims and swoops over water. Mass in tremendous fall flocks and on wintering grounds.
- **ID:** Glossy blue-green back and pure white underparts. Females have drabber backs and young birds are brown-backed. Length 5½ inches.
- **Habitat:** Open areas near water, including coasts.
- **Nest:** In natural tree cavities, eaves of buildings, nest boxes.
- **Food:** Winged insects taken in flight. In fall, bayberries taken in large quantities.

bicolor

TREE SWALLOW

Gray Jay

Perisoreus

HIKING IN NORTHERN CONIFEROUS FORESTS among evergreens capped with puffs of snow, one often hears a clear, hollow, whistling sound — "*teelawoo wheeoooo chuck chuck.*" Pause, and you will soon be investigated by the most inquisitive of the northwoods' birds, the Gray Jay. It looks like an overgrown chickadee with muted grays and whites and seems more fluff than substance. With wide wings quickly flicking in and out, they glide roller-coaster fashion from tree to tree as they approach. At nearly a foot long, they are impressive birds and will come in very close to inspect an intruder and, more importantly, to check if there is any food available. They show no shyness. I have had birds hop onto my shoulder and literally pluck pieces of bread from my hand. This tameness is not shown by any other North American bird.

Juveniles are deep sooty colored with a bright-yellow mouth gape. Always begging for food, they follow the foraging flock with hope of effortlessly obtaining a meal.

ATTRACTING THEM TO YOUR GARDEN

This bold, inquisitive bird lives in woods and often visits campsites for food scraps. They will come to beef suet and food scraps. Smear bacon fat or peanut butter into pine cones and they will spend time cleaning it out.

DISTRIBUTION

Boreal forest from Alaska to Newfoundland south to Canada and the north of the U.S., as well as in the Rockies and south Cascades.

THE SONG

teela woo wheeoo chuck chuck

Note sequence: Bubbling, guttural, and warbled notes often delivered on the wing; also a harsh chuck chuck.

Time of song: All day.

Other birds with similar song: Steller's Jay and Clark's Nutcracker also "chuckle."

IN BRIEF

Behavior: Extremely tame; roams about in loose, family flocks; has a very buoyant flight.

ID: Fluffy gray and white plumage, short bill, and a fairly long tail. Young birds have sooty gray plumage and a pale bill. Length 11½ inches.

Habitat: Northern coniferous woods.

Nest: A bulky mass of twigs and grasses lined with moss; on a branch near trunk of evergreen.

Food: Omnivorous including the young of other birds; makes caches of food; at feeder will take seeds, bread, cracked corn, and suet.

canadensis

Cyanocitta cristata

Blue Jay

Cyanocitta

THE BLUE JAY is a familiar species in parks, backyards, even in the trees that line the streets of the largest cities, and its antics are often a source of dismay. At the feeder they bully and fill their mouths to overflowing. They take this hoarded loot off to an old nest or other site and hide it for future ingestion. Afield, they are notorious nest robbers, often cleaning out the young of all the birds nesting in their vicinity. In addition, they are the alert system for the woodlands. All you need to do is enter an area with jays, and the alert call is given for other birds to hear no matter how quiet you are.

Their vocal repertoire is amazing. From low guttural notes that one can hear but a few feet away (called a whisper song), to the loud screams of "*jay–jay–jay*" that give them their name. They call like red-shouldered hawks, flickers, goshawks, and other birds of their area. Thrown in are sounds from a frog croaking to the familiar clothesline being reeled in. And they are beautiful — too often a bird such as the jay is so common that we consider it a pest species and fail to take time to admire its beauty.

ATTRACTING THEM TO YOUR GARDEN

The Blue Jay is a bold, aggressive bird that will plunder other birds' nests, pester hawks and owls, and even chase a cat. It needs no attracting to the birdfeeder as it will eat anything it finds there and will always find a way to get at the food.

DISTRIBUTION

Across the eastern two thirds of the U.S. and southern Canada.

THE SONG

jay-jay-jay tweelee tweelee tweelee

- **Note sequence:** Besides the loud, harsh jay-jay it has a wide range of notes, soft whisper songs of low whistles, loud mimicry calls of hawks, and two-parted explosive note sequences.

- **Time of song:** All day.

- **Other birds with similar song:** American Goldfinch (*Carduelis tristis*), see page 228.

IN BRIEF

- **Behavior:** Loud, bold; moves about in groups.

- **ID:** Brilliant blue with white spotting and black necklace. The female is marked identically except for white wing markings. Length 11 inches.

- **Habitat:** Forest, woodland (especialy with oaks), orchards, parks, gardens.

- **Nest:** A bulky cup of sticks, grapevine, in limb crotch of oak, maple, or large shrub.

- **Food:** Omnivorous, especially acorns and beechnuts; readily comes to feeder.

Corvus brachyrhynchos

American Crow

Corvus

THE CROW'S ABILITY TO LEARN is legendary and studies have shown that they are among the most intelligent of all bird species. Besides its reputation as a "villain," it is also known for its intricate behavioral patterns. It is one species that almost seems to be able to think through a situation and weigh the outcome. It is a very social species and roosts of thousands are not uncommon. When foraging for food, working parties scour the area while one bird is always on the lookout for danger. Every aspect of wariness and cunning is used in obtaining food, and being an omnivore, anything is taken. Be it stealing from a smaller bird, getting food from a dumpsite, grabbing unattended food, or actually working to find it, it is certain that most of the time the crow will meet with success.

Though intelligent, there is no truth that splitting a crow's tongue will allow it to talk. They can mimic sounds and words and have a wide repertoire of gurgles, harsh yells, and bubbling sounds in addition to the familiar "*caw caw caw.*"

ATTRACTING THEM TO YOUR GARDEN

As crows forage, they are sure to take any food available at your feeder, from seeds and berries to food scraps. If oaks are in your yard, they will take the acorns. All fruiting trees will also be used as feeding sites.

DISTRIBUTION

Ranges across Canada and the U.S., coast to coast. Only pockets of the southeast lack this species.

THE SONG

caw-caw-caw caw-caw-caw

- **Note sequence:** Loud and distinctive, rapid notes. Also guttural croaks and a complex note sequence.

- **Time of song:** All day.

- **Other birds with similar song:** The Fish Crow has a duck-like quack; the Raven "croaks."

IN BRIEF

- **Behavior:** Moves in organized groups; often roosts in tremendous numbers. Very adaptive.

- **ID:** Well-known for its large size and black color. Length 18 inches.

- **Habitat:** Open woodland, parks, farmlands, rangelands.

- **Nest:** A bulky mass of sticks in crotch of tree.

- **Food:** Omnivorous.

brachyrhynchos

Black-capped Chickadee

To the backyard birdwatcher, the Black-capped Chickadee is the life of the feeding station or tray. Their rollicking call of "*chick–a–dee–dee*" signifies their presence often before they are seen. They often forage with close "relatives" such as the Tufted Titmouse or the White-breasted Nuthatch, making up a very effective hunting party. In the fall, they will associate with the moving groups of migrant warblers and thereby alert the birder to the presence of birds that might otherwise be overlooked.

In the spring, their true song is heard — a drawn-out "*feee-beeee.*" For the uninitiated, it sounds like it should be a phoebe, but it is the chickadee on its territory or looking for a mate. The nest, a natural cavity, may produce six young that actively forage with the adults within a few weeks of hatching. They will nest in nestboxes placed at the wood edge.

ATTRACTING THEM TO YOUR GARDEN

The chickadee is a familiar visitor to the birdfeeder, and will also eat berries and insects garnered from the branches and trunks of trees. With consistency of placing out food, they quickly learn the routine and arrive at the same time each day to "walk to the feeder with you." Their inquisitiveness and trust make them a species that is easy to attract to the outstretched hand with food.

DISTRIBUTION

Alaska to Nova Scotia and south of Colorado and Virginia.

THE SONG

fee-bee fee-bee/chick-a-dee-dee

- **Note sequence:** A plaintive two-parted fee-bee dropping in pitch at the end. The better-known call is a distinctive and lively chick-a-dee-dee.

- **Time of song:** All day.

- **Other birds with similar song:** The Carolina Chickadee has a longer call with more phrases.

IN BRIEF

- **Behavior:** Moves in groups; very active, hangs from branch tips or flutter feeds under leaves.

- **ID:** Black cap and bib. Length 5 inches.

- **Habitat:** Mixed forest, pine woods, swamplands, parks, orchards, backyards.

- **Nest:** Old woodpecker holes, natural cavities or excavates own holes in soft wood.

- **Food:** Insects and invertebrates in summer. Winter diet of seeds and berries.

BLACK-CAPPED CHICKADEE

Poecile hudsonicus

Boreal Chickadee

Poecile

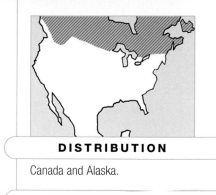

CANADA AND ALASKA, with their extensive boreal forests, play host to this species year-round. In the lower 48 states, outside of northern New England and the Cascades of the Northwest, it remains a visitor from time to time. In some years, irruptions carry scattered birds or small groups far south of their normal extension.

In appearance it is similar to the well-known Black-capped Chickadee except that the top of the head is brown and the sides are a frosted tawny color. The call is a distinct "*seek-a-day-day*" with a strong accent on the ending syllables. As with all members of the chickadee complex, they love to wander about in loose flocks investigating every nook and cranny of trees' bark and foliage.

ATTRACTING THEM TO YOUR GARDEN
Because of its desire to stay within the evergreens, the Boreal Chickadee is rarely encountered in a garden situation unless it lies within the boreal area. Under these circumstances they will come to a feeder for peanut hearts, peanut butter, and sunflower seeds. In the lower 48 states when it puts in an appearance at a feeder, it may stay for the winter's duration. With the onset of spring, they will join the movement of chickadee flocks northward and return to their evergreen world.

DISTRIBUTION
Canada and Alaska.

THE SONG

seek-a-day-day   see-you

- **Note sequence:** Call is repetitious with accented rise at the end; song is a repeated warble.

- **Time of song:** All day.

- **Other birds with similar song:** The call of the Black-capped Chickadee (*Poecile atricapillus*), see page 78, is less accented and emphatic.

IN BRIEF

- **Behavior:** Roams in small flocks; hangs from limb tips.

- **ID:** Brown cap, brown back, white cheeks, and brown flanks. Length 5½ inches.

- **Habitat:** Northern coniferous forests.

- **Nest:** Natural cavities, old woodpecker holes, excavates own holes in soft wood.

- **Food:** Insects, their eggs and larvae, and spiders.

Tufted Titmouse

A CALL LIKE A PERSON whistling for their dog "*wheeee-o-weet*," or a loud "*peter–peter–peter*" in rapid succession indicates the presence of this bird. It is one of those species that one invariably hears before seeing it. When located, it is rarely alone but in the company of other titmice or their close relatives, the Black-capped Chickadee and White-breasted Nuthatch. The party moves rapidly through the trees covering every nook and cranny in search of food.

Upland woods, evergreen forests, lowland swamps, backyards, and formal gardens all are the haunts of this gray little juggernaut. When finally seen, the beady little black eye stands out below the distinct crest. In Texas, the form called the Black-crested Titmouse occurs with a white forehead and a coal-black crest. A permanent resident throughout their range, they have slowly expanded northward over the last 30 years. One factor adding to their expansion has undoubtedly been feeding stations. In the winter, populations take up residence near feeders and will be daily visitors.

ATTRACTING THEM TO YOUR GARDEN

The Tufted Titmouse is inquisitive, relatively tame, and a frequent visitor to the birdfeeder. It will also take readily to nesting boxes placed at the wood edge or within the wooded part of a large garden. They love sunflower seeds and will often take them from your hand.

DISTRIBUTION

Eastern half of U.S., range expanding.

THE SONG

peter-peter-peter

- **Note sequence:** A loud and ringing staccato series, with loud whistles.
- **Time of song:** All day.
- **Other birds with similar song:** Parts of song can be mistaken for chickadee notes.

IN BRIEF

- **Behavior:** Active, inquisitive, moves in groups; a frequent feeder visitor.
- **ID:** A gray bird with a head crest, dark button eyes, and orange-tinted sides. Length 6 inches.
- **Habitat:** Woodlands, swamplands, orchards, parks, backyards.
- **Nest:** A hole nester in natural cavity or birdbox, nest is lined with leaves. Nest boxes will be used.
- **Food:** Seventy percent of diet is insects, also seeds and berries. At feeder, favorites are peanut butter and seeds.

bicolor

TUFTED TITMOUSE

Oak Titmouse

Baeolophus

IN THE OAK WOODLANDS OF coastal California, a distinctive *"tick-see-day-day"* call often leads the East Coast birder in pursuit of what they believe will be a chickadee. In its place will be found a plain, drab brownish titmouse. Resident and fairly common within these oak woodlands, they seem to prefer to work in groups, at the bases of trees, or very often hopping about on the ground itself. In the spring, as the brown hillsides of California come into bloom and begin greening, the loud *"weety weety weety"* song can be heard echoing through the woods. A similar species (the Juniper Titmouse) lives on the other side of the Sierra Nevada, and is an inhabitant of juniper and pinyon pine forests. The call of this species is nearly the same, but the color is a lighter mousey gray.

As with the rest of the family, this titmouse is very active and moves about in small bands with other related species. They take great joy in the discovery of a predator, such as a western screech owl or pygmy owl and will scold and flit about it for some time. These discoveries are often made as they are taking food to hide in some tree crevice or cavity.

ATTRACTING THEM TO YOUR GARDEN

At the feeder, the Oak Titmouse will take peanut butter and sunflower seeds. Be sure water is available. If you are resident to an oak wood area, they will use nest boxes.

DISTRIBUTION

Dry oak woodlands of California; just reaching Oregon.

THE SONG

tick-see-day-day weety weety weety

- **Note sequence:** Wheezy and accented dropping in pitch at the end.

- **Time of song:** All day.

- **Other birds with similar song:** Can be confused with the song of Black-capped Chickadee (*Poecile atricapillus*), see page 78.

IN BRIEF

- **Behavior:** Very active; moves in groups, skulks in thickets.

- **ID:** Gray bird with a small crest. Length 5¼ inches.

- **Habitat:** Oak, juniper and pine woodlands, and scrub.

- **Nest:** A hole nester including fence posts.

- **Food:** Small insects, seeds and berries; at feeder takes peanut butter and sunflower seeds.

Sitta canadensis

Red-breasted Nuthatch

Sitta

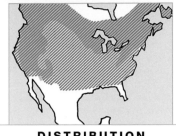

MANY BIRDS ARE NAMED after what they do. In this case, "hatch" refers to the opening of nuts and seeds of all types, rather than the opening of eggs. At this job, nuthatches are quite adept. For a small bird, many seeds and fruit types can be formidable tasks to open. The seed is usually wedged into a crevice and pounded on until it is "hatched."

Within evergreen woods, the Red-breasted Nuthatches work over every trunk and limb, searching crevices and opening cones for egg masses, larvae, seeds, and small insects. The common call is a high-pitched nasal "*yank*."

This species is an irruptive migrant. On wintering grounds in some years, there is a great paucity of the species and then years follow when the evergreen woods are full of them. Like other irruptive species, this can lead to nesting in areas such as woodland edges and backyards where they normally would only visit. Nesting can be in either deciduous or evergreen trees.

ATTRACTING THEM TO YOUR GARDEN

These birds will take to nest boxes, which should be placed as close to a prime habitat as possible because they are not as adaptive to interacting with man as the White-breasted Nuthatch. They come readily to feeding areas during the winter months, taking suet and sunflower seeds with great relish.

DISTRIBUTION

Across Canada and the northern portion of the U.S., moving south in the winter.

THE SONG

yank yank yank yank yank yank

Note sequence: A nasal series of rapid notes.

Time of song: All day.

Other birds with similar song: The White-breasted Nuthatch *(Sitta carolinensis)*, see page 88, has a similar but lower, less emphatic call.

IN BRIEF

Behavior: Works limbs and trunks of trees in head-down position; gregarious, joins chickadees, etc. in foraging position.

ID: Blue-gray back, black cap, white eye line, black line through the eye, and rusty underparts. Length 4½ inches.

Habitat: Mainly coniferous woods but into mixed woodlands especially during irruption years.

Nest: Excavates natural cavities in soft wood, old woodpecker holes; in conifer, surrounds entrance hole with resin.

Food: Small insects, their egg cases and larvae, pine seeds, mixed berries.

canadensis

Sitta carolinensis

White-breasted Nuthatch

Sitta

A NASAL CALL "*yank-yank-yank*" is sure to attract attention and the beginning birder who locates the bird has the surprise of seeing a bird walking head-first down a tree! Such diagnostic behavior makes it easy to identify. If one looks closely at the bill, it can be seen to taper to a chisel point. It is the use of this bill to hatch or open nuts and other seeds that has given this bird its name. To open seeds, it will hold them with its feet or take them to a rough surface such as tree bark, wedge the seeds in, and hammer them open. If you have a feeder it is not uncommon to hear pounding on the side of the house early in the morning as a bird jams the seeds in between shingles and then pounds them to split the hard coat. When a good food source is found such as a full feeding station, caching food is another habit. Sunflower seed after sunflower seed will be taken away and stored.

The song is along the same lines as the call, a series of nasal whistles on a non-melodic uniform pitch.

ATTRACTING THEM TO YOUR GARDEN

The White-breasted Nuthatch is a frequent visitor to birdfeeders, where sunflower seeds are a favorite. It prefers wooded areas but has adapted well to city and suburban environments. Birdboxes are also readily accepted and will be used in the winter as a night roosting area.

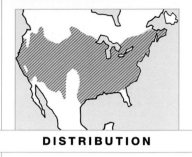

DISTRIBUTION

Across the U.S.

THE SONG

yank-yank-yank hit-hit-hit

- **Note sequence:** The call is a staccato series of rapid nasal notes; the song is short and high-pitched. (There is also a low-pitched series.)

- **Time of song:** All day.

- **Other birds with similar song:** The Red-breasted Nuthatch *(Sitta canadensis)*, see page 86, has a more rapid and more nasal call of the same pattern.

IN BRIEF

- **Behavior:** Creeps head downwards down tree trunk.

- **ID:** Jet-black cap (gray in females), pure white underparts and rust-red flanks. Length 5½ inches.

- **Habitat:** Woodland, pines, parks, gardens.

- **Nest:** In tree cavity or nest box.

- **Food:** Insects, spiders, seeds, and berries; at feeder, sunflower seeds are a favorite.

Certhia familiaris

Brown Creeper

Certhia

WHILE WALKING UPLAND WOODLANDS, a birder may be drawn by a high-pitched "*seeeet*" note and his or her eye attracted to what seems like a piece of bark or lichen moving up the trunk of a tree! As the bird pauses to glean an egg case or insect larvae from a fissure in the bark, its actual form can be made out — slim with brown and buff streaks, extremely well-camouflaged for creeping up trees. The bill is fairly long and curved, just right for probing into cracks and crevices. The feet are unseen because the bird hugs the bark so tightly, and the tail, a rusty color, is sharp-tipped and used as an effective prop as the bird works its way up a trunk to obtain a prey item.

The Brown Creeper is enjoyable to watch in all of its habits. It spirals up a tree, seemingly in quick short "jumps" and upon reaching the uppermost area of investigation, simply drops off one tree and flits to the base of another to start a new upward journey. In courtship it has a wonderful, descending musical jumble of notes.

ATTRACTING THEM TO YOUR GARDEN

During migration and winter months, they will come to feeders for beef suet or suet seed blocks. Be sure suet is wedged in the tree as creepers seldom leave the trunk for foraging.

DISTRIBUTION

Across the U.S. and south Canada.

THE SONG

seet-tweedlee-deet

- **Note sequence:** A high-pitched short series, delivered very rapidly.

- **Time of song:** Morning and evening.

- **Other birds with similar song:** Similar to the trilled song of Golden-crowned Kinglet but more musical.

IN BRIEF

- **Behavior:** Creeps up trees. Flies down to the base of another tree and works up trunk in a spiral.

- **ID:** Brown and buff camouflage and rusty tail. Length 5 inches.

- **Habitat:** Mixed woods and pines.

- **Nest:** Grasses and sticks wedged behind loose bark strips.

- **Food:** Gleans bark for small insects, their egg cases, and larvae, also spiders and a few seeds; at feeder will take suet.

familiaris

Canyon Wren

Catherpes

STANDING AT THE BASE of canyon walls that seem to rise so high as to narrow the opening at the top to a mere slit of blue sky, the echoing song of the wren seems ethereal. Rapid downwardly spiralling notes slow in speed towards the ending: *"tee–too, tew, tee–too–tee."*

Finding this songster can be a daunting task. It scampers in and out of cracks and crevices like a mouse. Even when it stops to peer down at an intruder, all that is usually seen is a head quickly popping over a rock edge, and then quickly gone. When finally viewed in the open, creeping up a rock face as it probes for insects, eggs, and larvae, the rich chestnut of the back shows in striking contrast to the pure white throat. Finding a morsel, it drops from the face as if shot and disappears into the pile of rocky rubble at the base. The walls and roofs of cabins set within these canyons are also eagerly investigated in the quest for spiders and insects.

The nest of twigs, stems, and leaves is wedged into a crevice of the rock face.

ATTRACTING THEM TO YOUR GARDEN

In western gardens, may nest near houses and under crevices in buildings. Comes to gardens to forage in stick piles and will take suet and some seeds.

DISTRIBUTION

Western U.S. from central Washington south through California and west Texas to Mexico.

THE SONG

tee – too tew tew too tew

Note sequence: A loud ringing series of liquid notes which spiral downwards.

Time of song: All day.

Other birds with similar song: The single "jeet" note can be similar to the Rock Wren of the same habitat.

IN BRIEF

Behavior: Creeps about mouse-like on rock faces. Will peer at an intruder then quickly disappear.

ID: White throat and breast and chestnut belly. Long, thin bill. Rust tail crossed by black bars. Head, back, wings, and abdomen barred and spotted with black and white. Length 5½ inches.

Habitat: Steep-sided canyon faces and rock slides.

Nest: Bulky mass of sticks lined with hair or fur, wedged into a rock crevice.

Food: All forms of insects, spiders.

mexicanus

CANYON WREN

Cactus Wren

Arid scrub and cactus country is the home of this large and attractive wren. As it hops up through the cactus blades and assumes its posture at its singing perch, it takes on the familiar erect posture with drooping tail of many of the wrens. Head thrown back and entire body shaking, it pours forth a barrage of "*cha-cha-cha-cha-cha*" notes interspersed with a few harsh warbling notes. Upon completion, the Cactus Wren usually slips back into the thicket and its presence is known only from an occasional harsh scolding "*ka-ka-ka*."

The nest is often placed in its favorite, the very spiny cholla cactus, or in one of the prickly pears. In one cactus clump there may be many such nests; one to be used for nesting, others as roosting sites. These cacti areas are often frequented by snakes, and battles between many Cactus Wrens and intruding snakes in search of eggs and young birds can cause a massive uproar in an often serene desert setting.

ATTRACTING THEM TO YOUR GARDEN

During the year, especially after the nesting and raising of young is over, the birds will move in closer to civilization. They will visit feeders, and one favorite object is half an orange stuck on a post. The birds will sit by the hour plunging their long curved bill into the fruit, allowing the viewer plenty of time to study their rich brown, tawny, and boldly streaked body patterns.

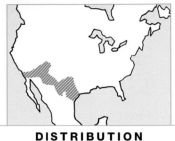

DISTRIBUTION

Southwestern U.S. from southern California to south Texas on into Mexico.

THE SONG

cha cha cha rach chick ra ra

Note sequence: Scolding series of harsh notes rising and falling in sequence. Also a rollicking, gurgling call.

Time of song: All day.

Other birds with similar song: All species of wrens tend to have harsh, scolding notes.

IN BRIEF

Behavior: A skulker, but sings from conspicuous position; attracted by "squeaking."

ID: Broad white eye stripes, heavily spotted breast and barred wings and tail. Length 8½ inches.

Habitat: Dry areas with cacti.

Nest: Globular mass of grasses and twigs with side entrance, sited in a cactus.

Food: Insects, cacti fruit, at feeder attracted by fruits such as half oranges.

Bewick's Wren

Thryomanes

ONE INTERESTING ASPECT of studying birds is determining why certain species are found where they are. It is obvious that birds that occupy the same niche or lifestyle cannot effectively co-exist in the same area. Yet, in Bewick's Wren we have a bird that appears to lead a double life, depending on the area studied. In the West it is common, interacting with the less common House Wren, and appears to be filling the role of the Carolina Wren of the east. It prefers the dry chaparral thickets that the House Wren seems to disdain, and so in such areas there is no conflict. In areas where the Carolina is dominant, the House Wren is basically missing and Bewick's takes on the House Wren niche. And in areas where all three are present, the Bewick's appears to be losing ground and withdrawing its range to more stable areas.

It has typical wren actions: very quick, tail cocked high, and a persistent singer. The song is a series of clear introductory notes that seem to be delivered as if inhaled and then exploded, followed by a falling series of trills.

ATTRACTING THEM TO YOUR GARDEN

At the feeder, the Bewick's Wren will take seeds, suet, and fruit. They will also make use of birdboxes for nesting. Provide brush piles for foraging and a water supply.

DISTRIBUTION

Western Texas south to Mexico and west to California. North on the coast to Washington with pockets in the southeast.

THE SONG

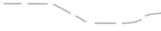

seep seep – jejeje jeet jeet

- **Note sequence:** A jumbled series of chattering notes starting with clear whistles which sound as if they are inhaled.

- **Time of song:** All day.

- **Other birds with similar song:** Similar to House Wren *(Troglodytes aedon)*, see page 98, but raspy "inhale" note at start distinguishes it.

IN BRIEF

- **Behavior:** A thicket skulker, and inquisitive, it will pop up and peer at an intruder.

- **ID:** Brown back, white line over the eye, and gray underparts. Long tail fringed with white. Length 5¼ inches.

- **Habitat:** Thickets and waste areas, overgrown fields, gardens, chaparral of the West Coast.

- **Nest:** Bulky nest of twigs and grasses in cavity.

- **Food:** Insects, their egg cases, and larvae; at feeder takes seeds, suet, and fruit.

bewickii

BEWICK'S WREN

House Wren

Troglodytes

FEW BIRDS HAVE ADAPTED better to alterations of habitat than the House Wren, which has taken advantage of man-made structures and waste. Birdhouses are readily accepted for a home but so is the eave corner of a garage or old shed. The nest is a bulky structure of coarse sticks collected new or re-used after being meticulously removed from a nest box by the returning male. Even though it is the same nest used the previous year, it will be removed and rebuilt. To thwart competition for food within its territory by other species, the House Wren also has the unlikeable habit of piercing other birds' eggs with its longish bill.

Its song is an explosion of sound. Roller-coasting up and down, the jumble of gurgling and bubbly notes seems to pour out and roll on and on. After the nesting season, it falls silent and assumes rather secretive ways. It can be difficult to see at this time.

ATTRACTING THEM TO YOUR GARDEN

House Wrens are easily attracted to nest boxes in the garden. At least two boxes should be put up because wrens like to build dummy nests to fool predators. One may also be used as the night roost box by the male. Make a slot at the side of the entry hole as this allows the birds to maneuver longer twigs. Wren boxes should be at least four feet off the ground.

DISTRIBUTION

Across south Canada and throughout the U.S.

THE SONG

tiddle a see-che che che che tiddle lee

- **Note sequence:** Rapid series of raspy, harsh notes and a bubbly, explosive sequence.

- **Time of song:** All day.

- **Other birds with similar song:** Most wrens have bubbly songs and wheezy scolds.

IN BRIEF

- **Behavior:** Skulker of brush piles, very active, holds tail cocked up.

- **ID:** Paler brown below than above with obscure flank bars. Pale line over the eye. The tail is often cocked at an angle but when working in the thicket is held straight out. Length 4½ inches.

- **Habitat:** Woods, parks, gardens, backyards, scrubby hillsides.

- **Nest:** Bulky mass of sticks placed in tree cavity or nest box. Several dummy nests may be built.

- **Food:** Insects, their larvae, and egg cases; also spiders.

aedon

HOUSE WREN

Winter Wren

Troglodytes

THIS TINY MITE of a bird often slips by unnoticed, especially during migration when it travels from the northern extremes of its range to its southern limits. It tends to be shy and its size makes it difficult to locate even when the birder knows it is creeping about in a nearby thicket. More often than not, the sharp *"chip-ship"* note in rapid sequence gives its presence away. Skulking around near or on the ground, it may pop into view for a moment, make brief scolding noises, then slip back into the dense thickets. During the breeding season, its preferred habitat is the cool, damp, coniferous forests of the north and west. The sweet, twittering, endless song of this small bird is heard to best advantage here.

The nest is placed on the ground, under an overhanging edge, rocky concavity, or in a dense brushy area.

For the majority of American birders, the name "Winter Wren" holds true as it is most often seen on migration or during the winter months. It uses a wide range of habitats, but is always around a dense thicket or brush pile.

ATTRACTING THEM TO YOUR GARDEN

Perhaps the best attractant is a well-made brush pile at the garden edge to allow hunting for insects. They will on occasion come to a suet feeder.

DISTRIBUTION

Across south Canada and in New England and the Pacific northwest to California and south in the Appalachians.

THE SONG

chip-ship

- **Note sequence:** A beautiful wild ringing trill made up of two-parted notes; call is short and sharp.

- **Time of song:** All day.

- **Other birds with similar song:** The chip note is typically wren in quality, but the song is unique.

IN BRIEF

- **Behavior:** Very active, bobs up and down, skulks about thickets. Secretive and hard to spot.

- **ID:** Deep brown with heavy barring on the flanks. This bird has a short stubby tail, which it cocks up. It also has an unusual bobbing movement. Length 4 inches.

- **Habitat:** Mountain ravines, dense coniferous forests, coastal thickets, parks, and gardens.

- **Nest:** Domed structure of rootlets, plants, fibers, and moss.

- **Food:** Insects and other invertebrates.

troglodytes

Cistothorus palustris

Marsh Wren

Cistothorus

SITTING QUIETLY AT THE EDGE of a cattail or reed marsh in a fresh or brackish situation, you will hear the loud, rapid, jumbled notes of the singing male Marsh Wrens. Small tawny brown forms will be seen hurtling over the cattails, then plunging from sight. To see one in close, make a squeaking sound on the back of your hand and wait. Soon a scolding "*sh-sh-sh-sh*" will be heard followed by "*check, check, check*." At last the wren will come into view, usually hanging spread-legged between two cattail stalks. Tiny in body, the strong white line over the eye accentuates its staring inquisitiveness.

Nests are football-shaped and are suspended from plant stems. One bird may build 15 nests on its territory; one for nesting, usually the most difficult to locate, and the others to act as decoys for predators, therefore in much more obvious positions.

Food consists of all forms of small insects and invertebrates. Singing will go non-stop on through the nesting season. As the marsh begins to freeze in late fall, most birds head south, but some linger on in the north and manage to find the energy to survive.

ATTRACTING THEM TO YOUR GARDEN

The only chance to see this wren in a garden situation is if you live near a marsh edge and enhance a portion of your land by flooding, or if you have a pool and plant dense cattails.

DISTRIBUTION

Across northern half of U.S. and western Canada, coastal California, Nevada and Utah.

THE SONG

sh-sh-sh

- **Note sequence:** A bubbling series of harsh notes delivered in a rapid rolling pattern; call sh-sh-sh or check-check.

- **Time of song:** All day.

- **Other birds with similar song:** Sedge Wren has bubbly, rattling song.

IN BRIEF

- **Behavior:** Climbs nimbly about reeds, sings from reed stalk; flies with stiff wings.

- **ID:** Solid rust-brown cap and streaked back. White above the eye and on the underparts. Length 5 inches.

- **Habitat:** Fresh and brackish reed and cattail marshes.

- **Nest:** A football-shaped mass of plant fibers with side entry hole. Builds dummy nests to fool predators.

- **Food:** Small insects and invertebrates, seeds on occasion.

palustris

MARSH WREN 103

Cinclus mexicanus

Dipper

Cinclus

ALONG THE CLEAR RUSHING TORRENTS of mountain streams of the western United States, the birder will often catch sight of a uniform gray ball-like bird "buzzing" up- or downstream on fast beating wings. As it passes, a sharp alarm call "*beezeeep*" can be heard. The bird is the Dipper or Water Ouzel.

The Dipper's beautiful, strong song is made up of trills and flute-like notes more like that of a thrush or wren than a Dipper. If persistent, the birder can often find the Dipper working the rocks along the river's edge and this is when it is seen to best advantage. As it bobs along from rock to rock, it is constantly looking for invertebrates. Finding a shallow edge to its liking, it walks directly in and slowly disappears from sight. Observing such a pool from above, the bird can be seen working its way along the pool bottom. Upon securing food it bobs to the surface and quickly takes flight to the nest site.

Dippers seem to love to sing, and even through the winter where streams stay open, the songs can be heard cascading down the riverine canyons.

ATTRACTING THEM TO YOUR GARDEN

The only chance to see them on your property is if you have a large, torrent stream running through it or a waterfall to allow nesting nearby. No attractant food can be placed out.

DISTRIBUTION

Western United States.

THE SONG

see-ta si si si la te lasee see

- **Note sequence:** A fluid, bubbling flow of two main phrases, the first on a higher pitch.

- **Time of song:** Morning and evening.

- **Other birds with similar song:** Similar to the prolonged song of Winter Wren (*Troglodytes troglodytes*), see page 100, but much louder and more forceful.

IN BRIEF

- **Behavior:** Bobs as it walks by edge of streams.

- **ID:** Stout body and short tail. Gray body and brownish head and neck. Length 7½ inches.

- **Habitat:** Rocky river and mountain streams.

- **Nest:** A well-woven ball of mosses with tunneled side entrance under waterfall or wet, dripping moss cliff face.

- **Food:** Insects, invertebrates, small fish.

mexicanus

Regulus calendula

Ruby-crowned Kinglet

"Fı-JIT-JIT-JIT" IS A COMMON SOUND in thickets during spring and fall migration across the continent. The caller is the Ruby-crowned Kinglet, one of the smallest North American birds. Hardly ever at a standstill, it flicks about in the bushes and has the constant habit of twitching its wings. Nervously moving about, it will often move in very close to the birder before working off through the thickets again.

Don't expect to see the ruby cap very often. Only when in courtship display (which involves flitting about with feathers raised, wings spread, and tail cocked) or when angry at a predator do the birds raise their vermillion central crown feathers into view. But once flared out, it is a sight not easily forgotten. Nor is the full song of the bird one that is difficult to forget. It is hard to believe that this mite of a bird has such a loud ringing song, an explosion — "*see-see-see teedle teedle teedle.*"

This species represents a group of immigrant birds known as the Old World warblers and have more direct ties with thrushes than our native wood warblers.

ATTRACTING THEM TO YOUR GARDEN

They pass through gardens and yards during migration. Brush piles and evergreen provide foraging areas. Water will also attract them. In winter, they may come to suet.

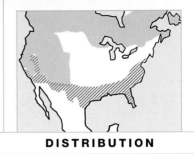

DISTRIBUTION

Alaska to Newfoundland south to mountains of the west and through the Appalachians in the east.

THE SONG

see-see-see teedle teedle teedle

- **Note sequence:** An explosion of notes following an introduction of several rapid, thin, high-pitched notes.

- **Time of song:** All day.

- **Other birds with similar song:** Call note is similar to Golden-crowned Kinglet.

IN BRIEF

- **Behavior:** Very active, always twitching wings out from body.

- **ID:** Grayish-olive with a distinct white broken eye-ring and white wing bars. Young fall birds have yellowish wing bars.The male has brilliant red brown feathers that it holds erect in display or when agitated. Length 4½ inches.

- **Habitat:** Prefers coniferous woodland for breeding, otherwise mixed woodlands and thickets.

- **Nest:** Tight cup of plant fibers lined with plant down and feathers and held in place by cobwebs.

- **Food:** Small insects, egg cases, spiders; some seeds and berries in winter.

calendula

Blue-gray Gnatcatcher

Polioptila

YOUR ATTENTION will probably first be drawn to the bird because of the thin, explosive *"spit-tseee"* it utters while working in a thicket or upper limbs of a tree. Some know the species as the "miniature mockingbird." Not only is the color pattern the same, but when it breaks into song, it is loud and rollicking.

Its favorite haunts are woodlands and thickets, and in the West it is found right down to the coastal scrub. The beautiful nest is constructed on the open ribs of trees. The young are tiny gray tufts of feathers that bounce along behind the adults begging food and quickly gobbling down larvae, eggs, and other insect material that is collected from bark crevices by the adults. The adults will often pursue insects in flight in the upper tree branches, twisting and turning with an audible clicking of the bill.

This long-tailed bundle of energy is a mite of a bird and is becoming more and more widespread to the north of its former range. At one time considered to be a southern species, the range has inched its way slowly northward to its present limitation in eastern Canada.

ATTRACTING THEM TO YOUR GARDEN

If you have large, lichen-covered oaks, this species may nest. Evergreens may also attract them to forage for insects.

DISTRIBUTION

Northeastern U.S. west to California and south to Mexico.

THE SONG

spit-see spit-see see see tiddleee see see

Note sequence: The call is high-pitched; the song a jubilant outpouring of lisps and warbled notes.

Time of song: All day.

Other birds with similar song: Quite unlike any other bird's call.

IN BRIEF

Behavior: Very active, inquisitive, runs about on limbs, flicking tail from side to side.

ID: Blue-gray with white eye-ring. Long tail edged with white with black central portion. Length 4½ inches.

Habitat: Mixed woodland, oak woods, chaparral, pinyon pine groves, dense hillside thickets.

Nest: Beautiful lichen-covered cup lined with plant down, looks like a knot on a limb.

Food: Mainly insects and other invertebrates.

caerulea   BLUE-GRAY GNATCATCHER

Eastern Bluebird

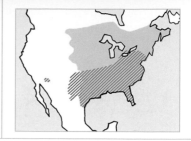

THE SIGHT OF AN EASTERN BLUEBIRD flying up from beside the road and perching on a fencepost is a sight that excites even the non-birder. The spectacular blue contrasting with the orange and white of the underparts is a striking combination. The song is a short, soft warble and is often repeated.

ATTRACTING THEM TO YOUR GARDEN

Most people are aware that they do not see as many bluebirds as they did in the past. Several things have affected the bluebird's nesting success. The introduction of the European Starling is a main factor. Eviction from nesting holes, be they man-made or natural, is commonplace. In addition, the species has very specific needs for a nest site, favorite sites being old trees with holes lining pastureland. But a great deal of the East has lost such pastureland and with it, the bluebird population. The birds can however be encouraged with birdboxes set on posts fairly low to the ground near a woodland edge, but in a position that is open on all sides. Unfortunately, such a site is also attractive to tree swallows which often usurp the nestbox. To counter this sabotage, "Bluebird Trails" have been established where large numbers of boxes are placed at intervals in ideal habitats, and in many areas the bluebird is making a remarkable comeback. At backyard feeding sites, they can be attracted to raisins, live mealworms, and even grape jelly.

DISTRIBUTION

Eastern half of the U.S.

THE SONG

ta ta tee o ta ta too lee

- **Note sequence:** A mellow series of notes usually sung at low level, sounds like a guttural warbling.

- **Time of song:** Morning and evening.

- **Other birds with similar song:** The Western Bluebird and Mountain Bluebird *(Sialia currucoides)*, see page 112, also have gentle, warbled songs.

IN BRIEF

- **Behavior:** Flutters about in open areas when feeding, often drops to ground to feed; has a very buoyant flight.

- **ID:** Blue upperparts contrasting with rich rust below and a white underbelly. Length 7 inches.

- **Habitat:** Open woodland edges, farmland, orchards, gardens.

- **Nest:** A cup of twigs placed in a natural cavity, abandoned nest hole, or nest box.

- **Food:** Mainly insects with some berries and seeds.

Sialis

EASTERN BLUEBIRD

Sialia currucoides

Mountain Bluebird

Sialia

AT THE EDGES OF rolling mountain meadows clothed in wildflowers or on the high gray-green sagebrush plains, dwells this bird of startling sky-blue color. In fact, the male Mountain Bluebird is entirely blue of various shades. In addition, it has a most peculiar feeding method. It forages for insects among low shrubs by hovering next to the branches and gleaning food items from limbs and leaves. With rapidly beating wings, it swings about the plant and upon grabbing an insect, will alight on the ground to eat it or flutter off on buoyant wingbeats to a nearby limb to feed. When first seen, it appears that someone is pulling a string, marionette-like, to keep the bird suspended and moving about the bushes.

The nesting site is a natural tree cavity or old woodpecker hole, often in aspen or birch trees. When winter comes to the high country, the population shifts to the rolling plains and extensive agricultural areas to the south.

ATTRACTING THEM TO YOUR GARDEN

Gardens in high mountain areas may see these birds visit for fruits of flowering trees and insects. Bird baths are also attractive. Leaving old tree snags with woodpecker holes may provide attractive nest sites.

DISTRIBUTION

Western North America from south British Columbia to Arizona and on to Mexico.

THE SONG

tur-lee tur-lee

Note sequence: Low series of warbled notes.

Time of song: Morning.

Other birds with similar song: Similar to Western Bluebird and Eastern Bluebird (*Sialia sialis*), see page 110.

IN BRIEF

Behavior: Forages by hovering low over the ground; masses in large groups in winter.

ID: Startling sky-blue color. Lacks the Eastern Bluebird's reddish-brown breast. Length 7¼ inches.

Habitat: Alpine meadows and mountain woodland.

Nest: In tree holes including old woodpecker nests; also nestboxes.

Food: Small insects and invertebrates.

currucoides

Veery

WITHIN THE EMERALD CONFINES of the moist mixed woods of the northern United States and southern Canada, a haunting flute-like song cascades down the scale. The songster is the Veery, a shy, reddish-brown thrush with faint chest spotting. To get to know this bird, one must venture into the moist woodlands and spend some time sitting and watching. It is not uncommon when one is perfectly still to see this thrush hopping about as it actively feeds among the leaves and mosses. At these times, it will approach quite closely.

The cup nest of bark, grass, and rootlets is often placed on the ground, but may also be tucked into the base of a shrub. The coloration of the eggs is a beautiful pastel blue that, in dark wet woodlands, appears to glow like embers from the nest.

The eastern birder has to be aware that when in the same habitat in the West, the songster may look different. A darker brown in color with more distinct spotting, it looks quite similar to its close relative, the Swainson's Thrush, but without the large eye-ring and pale sides.

ATTRACTING THEM TO YOUR GARDEN

As a passing migrant, it may seek cover in a well-planted garden where it can find deep shade. Water is also an attractant.

DISTRIBUTION

Across north U.S. and south Canada.

THE SONG

see to see tolee to wee

- **Note sequence:** A rolling series of descending notes that tumble downward with a flute-like quality.

- **Time of song:** Morning and evening.

- **Other birds with similar song:** In quality like a Wood Thrush (*Hylocichla mustelina*), see page 120, but it spirals downward.

IN BRIEF

- **Behavior:** Shy, hops about forest floor; has very erect posture.

- **ID:** Reddish-brown with faint chest spotting. Length 7 inches.

- **Habitat:** Wet uplands and mixed woods.

- **Nest:** Cup of grasses and fibers on ground or very low in shrub.

- **Food:** Flying insects such as dragonflies as well as beetles, grubs, slugs, and other invertebrates.

fuscescens

VEERY   115

Catharus ustulatus

Swainson's Thrush

Catharus

ALL THE NORTHERN FOREST thrushes seem to have the same basic color pattern, and this species is no exception. The rich olive to brownish color of the back is quite variable, and there is smudged spotting on the buff-tinted breast. Perhaps the best identifying mark lies in the wide buff-colored eye-ring, and in its habit of twitching its wings.

Swainson's Thrush is a bird of swamplands and moist woodland floors and in migration can be quite abundant. Indeed, on clear fall nights when the first cool winds cascade down from the North, hundreds of these birds migrating overhead in the blackness of night can be heard giving their diagnostic "*qweep*" calls. The song is whistled without the true flute-like quality of a wood thrush and is delivered in an ascending scale. It is usually sung from a low perch because the bird prefers to be close to the ground. When nesting, however, it breaks this rule, and I have occasionally seen the cup nest of twigs, leaves, bark strips, and lichen set as high as 20 feet up in a conifer. More often, it is placed a couple of feet off the ground in a scraggy shrub.

ATTRACTING THEM TO YOUR GARDEN

Seen mainly as a migrant in yards, shrubs for cover and water availability are lures for transient thrushes.

DISTRIBUTION

Alaska across Canada, south into the Rockies and northeast U.S.

THE SONG

tcc twaoo too oo ta loo

- **Note sequence:** A series of ascending whistles.

- **Time of song:** Morning and evening.

- **Other birds with similar song:** The Gray-checked Thrush song is very similar but spirals down at the end.

IN BRIEF

- **Behavior:** Often droops and twitches its wings; fairly shy — spends most of its time feeding on the ground but sings from treetop.

- **ID:** Light tan face and eye-ring. Lacks the red on the tail of the Hermit Thrush, and breast is more heavily spotted than the Veery's. Length 7 inches.

- **Habitat:** Cool northern forests or mixed woods on migration.

- **Nest:** Cup of grasses and plant fibers most often just above ground in shrub.

- **Food:** Wide variety of small insects, larvae, seeds, and berries.

ustulatus

Hermit Thrush

Catharus

All thrushes tend to have beautiful voices, but this species is often voted to have the best: a series of flute-like echoing notes that seem to drift in the air for seconds after the song is delivered.

The Hermit Thrush is a species of coniferous and mixed upland woods. Wet areas and cool ravine thickets are preferred. On the West Coast where the evergreen forests extend down to the shore, this thrush is fairly common, but its shyness gives no clue to its abundance. Across Canada, it is a widespread nester and on the East Coast, it nests in scattered localities of cool bog and swampland. In the South, it is a fall migrant and an overwintering species.

On the wintering rounds, its presence is often given away by a distinct "*chuck*" note as the bird forages in a dense thicket. If it does hop into view, it will raise its tail slowly to a cocked position, slowly lower it, and then flick it up again. This is an excellent field mark, as is the rusty tail and conspicuous white eye-ring. As with any species that inhabits a wide range of habitats, many subspecies occur but the reddish tail pattern is a consistent feature.

ATTRACTING THEM TO YOUR GARDEN

This thicket skulker appreciates brush piles and the shadows of low plantings. Wintering birds do come to suet on occasion and also to fruits such as raisins.

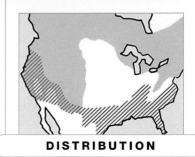

DISTRIBUTION

Across north North America from Alaska to Newfoundland and south in western and eastern mountains.

THE SONG

oo la la low/ah la la/ay i a la/la lee

- **Note sequence:** Pairs of flutelike notes rising and falling, ending with two-toned trills.

- **Time of song:** Morning and evening.

- **Other birds with similar song:** Wood Thrush (*Hylocichla mustelina*), see page 120, has a similar flute-like song but lacks prolonged trills.

IN BRIEF

- **Behavior:** Shy, sedentary; cocks tail as it hops about.

- **ID:** Brown back, white eye-ring, and rusty tail. Length 6-6¾ inches.

- **Habitat:** Evergreen forests, bogs, swamps.

- **Nest:** Compact cup of twigs, grass, and rootlets. Outside usually made of ferns and mosses.

- **Food:** Insects, worms, small snails, berries, and seeds. Has visited feeders on rare occasions.

guttatus

HERMIT THRUSH

119

Wood Thrush

THE WOOD THRUSH is most at home where the dappled light strikes the forest floor beneath a lush understory. Its habitat preference is wide: from laurel thickets of drier upland woods to the cool shaded swamplands of the eastern forests, the liquid notes of a rich, flute-like quality are a common summer sound. This thrush seems to prefer singing during the early morning, or evening hours just prior to sunset. Listen for the diagnostic "*pip-pip-pip*" call before the outpouring of the full "*ee-oh-lay*" notes that scale downward. These evening "vespers" will continue until just before dark when again the "*pip-pip-pip*" notes are delivered in rapid sequence until they taper off.

The nest, typical of the thrushes, is mud-lined. However, the wood thrush tends to add grass and rootlets to the inside, and the outside is loose in structure and gives the nest a rather shaggy appearance. It is placed firmly in the crotch of a shrub or small tree.

ATTRACTING THEM TO YOUR GARDEN

As a woodland species, they are reluctant to come to yards unless cover is available. Bird baths are often favored and they will come to orange halves at times.

DISTRIBUTION

Throughout eastern North America.

THE SONG

ee-oh-lay-ee-oh-lee

- **Note sequence:** A beautiful two-toned flute-like song.

- **Time of song:** Morning and evening.

- **Other birds with similar song:** Similar to a Hermit Thrush (*Catharus guttatus*), see page 118, but phrases are shorter and more spiraling.

IN BRIEF

- **Behavior:** Hops about on forest floor and low understory. Usually stays in shade and cover.

- **ID:** Reddish-brown head, warm brown back, and white underparts marked with dark teardrop spotting. Length 7¾ inches.

- **Habitat:** Prefers the cool understory near brooklets, swamp edges, and wet ravines.

- **Nest:** A mud-lined cup of grasses, plant fibers and paper, if sited near human habitation.

- **Food:** Insects, spiders, worms, some berries.

mustelina

Robin

Turdus

THIS IS THE BEST-KNOWN of all the songbirds in the United States. It can be found from the deep woodlands to the tiniest park in the largest of cities. During the early morning hours and late in the evening as sunset tints the treetops, the song of the Robin is heard to best advantage. Clear flowing notes see-saw back and forth in their emphasis. Liquid tones that are quickly learned by the birder are awaited by many to signal the end of a harsh northern winter.

In the spring when territorial boundaries are being established, the Robin can be seen singing from its favorite perches that delineate the area. During this time, it is also very defensive of the area and mad chases and tumbling fights often ensue when another male robin crosses its boundary. It is during this defense time that the males will also challenge their own reflections! Battles waged against windows and even hubcaps of cars will go on for hours.

ATTRACTING THEM TO YOUR GARDEN

Old apple orchards provide excellent overwintering areas for robins in the North. On the feeder, they will eat apple and orange halves. Care should be taken concerning lawn preparations to ensure against poisonings with toxic weedkillers. A flat, sheltered platform on the side of a building might be used for nesting.

DISTRIBUTION

Throughout North America.

THE SONG

tweedle lee taw dee to see to we

- **Note sequence:** A series of clear whistles with distinct pauses delivered in an up-and-down pattern.

- **Time of song:** Morning and evening.

- **Other birds with similar song:** Often mimicked by Mockingbirds (*Mimus polyglottos*), see page 128.

IN BRIEF

- **Behavior:** Hops about lawns; loves fruiting trees; large fall flocks.

- **ID:** A reddish breast extending to a white underbelly. Head is black with a broken white eyering. Back is a grayish-brown contrasting with the black tail that has white corner tips. Length 10 inches.

- **Habitat:** Everywhere! Fields, parks, orchards, backyards, mountain glades to coastal areas.

- **Nest:** A mud-lined cup of grasses and rootlets on a flat surface.

- **Food:** Earthworms, grubs, spiders; in fall, berries and fruit, will take seeds in winter.

migratorius

Varied Thrush

Ixoreus

WHEN I FIRST VISITED THE WEST and had the chance to walk through the magnificent cathedral stands of western evergreens, the haunting ethereal notes of the Varied Thrush impressed me as much as any song has. The distinct, nasal penetrating whistle was slow and drawn-out first on one pitch, then a second at a lower pitch, followed by a third higher on the scale. The songsters hopped into view atop a moss-covered stump. Their body conformation is almost identical to a Robin. The back a bluish gray, a bold orange line over the eye and the orange of the underparts shows a black chest band, and the wings are marked by orange wing bars which flash brightly when the bird is in flight. In general, these birds are quite shy and getting a good view is by chance or long vigilance.

It is a bird of the high mountain forests and the misty evergreen woodlands of the Northwest coastline. In Alaska, it can also be found in the alder woodlands of streambeds with dense tangles of Devil's club. In general a bird of the West, recently more and more records of these birds have been noted at winter feeding stations in the East.

ATTRACTING THEM TO YOUR GARDEN

Shrubs to provide cover and fruiting bushes and trees can prove attractive to wintering and migrant birds. When stragglers wander eastward in winter they will take seeds to make it through.

DISTRIBUTION

Western North America from Alaska to California.

THE SONG

wheeeee wheee wheee wheee

- **Note sequence:** A series of drawn-out buzzing whistles, each delivered on a different pitch.

- **Time of song:** Morning and evening.

- **Other birds with similar song:** A unique song unlike any other thrush.

IN BRIEF

- **Behavior:** Shy, feeds on forest floor; hops in robin fashion.

- **ID:** Male has orange underparts crossed by a black breast band, an orange line behind the eye, and orange-brown wing bars and patches. The female has a brown breast band and back. Length 9½ inches.

- **Habitat:** Dense coniferous forests.

- **Nest:** Cup of twigs, leaves, and grasses lined with mud and leaves, placed on limb near trunk.

- **Food:** Insects and earthworms.

naevius VARIED THRUSH

Townsend's Solitaire

Myadestes

THIS SPECIES LIVE in the high, cool, coniferous forests of the West. They are extremely sedentary birds. Perched on a dead branch usually quite low in the tree, they sit in a bolt-upright position surveying the area. Reluctant to fly except to capture food, they often trustingly allow the birder to pass by at a close distance. Their dull gray plumage affords them the camouflage to allow such bold indulgence. When they fly, however, distinct buffy patches can be seen in the wing, and the outer edge of the tail is white. From their perch on a snag or rocky outcrop, they plunge to the ground to snatch up an insect and then return to the roost site. The nest is a cup of mosses and twigs placed on the ground under a rocky overhang or the base of a shrub or tree.

The song is grosbeak-like in quality and long in duration (timed for over half a minute) and the birds often launch into flight during the song and fly in a circling pattern with wings aflutter. Upon the song's completion, the bird spirals down to the initial singing perch.

ATTRACTING THEM TO YOUR GARDEN

Moves through lower areas in migration. It may visit yards if fruiting trees are available. For the most part, a reclusive bird.

DISTRIBUTION

West coast from Alaska to Mexico through the Rockies.

THE SONG

too wee too were-too wee to see tee tweer

- **Note sequence:** Long in duration; sweet, clear notes often delivered in flight. Also a loud "yeek" note.

- **In flight:** Will sing in flight.

- **Time of song:** Morning and evening.

- **Other birds with similar song:** Some explosive notes of Flycatcher sound like "eek" note of this bird.

IN BRIEF

- **Behavior:** Very sedentary; flutters to ground from observation perch to feed.

- **ID:** At rest, brown-gray back and underparts and white eye-ring are distinguishing features. In flight, the tan wing patches and white outer tail feathers are visible. Length 8½ inches.

- **Habitat:** Cool northern evergreen forests, rocky hillsides, and ravines.

- **Nest:** Cup of mosses and twigs on the ground, usually under an overhang or in rock crevice.

- **Food:** Wide variety of insects, berries, and seeds.

townsendi

TOWNSEND'S SOLITAIRE

Mimus polyglottos

Mockingbird

Mimus

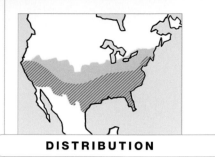

ON MOONLIGHT NIGHTS in early May from atop a dense thicket, the twisting, wing-fluttering form of the Mockingbird can be seen as it bursts forth with song through the night. One of the few songbirds that sing during the night, as its Latin name implies this is indeed a "many-tongued mimic." The species will mimic the local birds from every sector of the country in which they are found. As new species migrate in for the summer, it is not long before their songs are added to the Mocker's outpouring. The function of this mimicking has been a point of contention for a long time. Recent studies have shown that there is a positive effect to keeping birds of other species out of the Mocker's range via this song imitation, allowing less pressures on the Mocker's food supply.

The nest is tucked into the midst of the densest bush, which usually provides the main singing perch for the bird on territory. In the East, multiflora rose thickets are favored and the spread of the Mockingbird into the Northeast over the last 20 to 30 years is based on the spread of the rose.

ATTRACTING THEM TO YOUR GARDEN

Water, fruiting shrubs, and trees should attract this species. Rose bushes with small hips (the rose fruit) are an excellent food to get them through winter in the north.

DISTRIBUTION

New England across the lower Great Plains to California and south to Mexico.

THE SONG

rick chick chick wee

- **Note sequence:** Harsh paired calls rising and falling in a see-saw manner. During the spring, often sings all night.

- **Time of song:** All day.

- **Other birds with similar song:** Brown Thrasher *(Toxostoma rufum),* see page 132, sounds similar but always sings in double phrases.

IN BRIEF

- **Behavior:** Bold, often perches on top of brush.

- **ID:** Light and dark gray plumage. Length 11 inches.

- **Habitat:** Gardens, farmlands, parks, backyards. Any brushy areas. Multiflora rose is a key shrub.

- **Nest:** Large, loose structure of twigs, grasses, and rootlets placed in dense shrubbery. In desert areas, nests in cacti tangles.

- **Food:** Wild berries, especially rosehips, seeds, insects, and other invertebrates. Rare at feeder.

polyglottos

Dumetella carolinensis

Gray Catbird

THE CAT-LIKE MEWING coming from the thicket is the Gray Catbird. Its coloration is easy to pick out: all gray with a distinct black cap and a rusty undertail covert area. They are quite inquisitive and come in readily to any squeaking sound. At such times, they can be seen peering from the shrubs and bushes with their jet-black eye.

Catbirds belong to the family Mimidae, the mimics, which include the Mockingbird, Brown Thrasher and the like. When its song is heard, the harsh repetitive quality ties in with these relatives. It is not the mimic the others are, however, although within the jumbled notes of a song, some birds do interject some quite good mimicry.

ATTRACTING THEM TO YOUR GARDEN

Wherever there are dense shrubs and bushes, this bird will be found, from mixed woodlands to brushy field edges and the backyard brush pile.

It feasts on a wide variety of fruits and berries including blackberries, cherries, elderberries, cat brier berries, mulberries, and blueberries. Each year sees more and more catbirds remaining in the North for winter. At this time, watch them sneak to the feeder for suet and raisins set out through severe periods of weather. Those who stay on for the winter eat the fruits of holly, bittersweet, and honeysuckle. Seeds are taken only as an emergency food.

DISTRIBUTION

Eastern half of the U.S., lower Canada and spreading west to Colorado.

THE SONG

coyat eee ela toolee tee ta tee

Note sequence: A jumbled series of notes, often harsh and with prolonged lispy endings. Distinct "mew."

Time of song: All day.

Other birds with similar song: Similar to a Brown Thrasher *(Toxostoma rufum),* see page 132, but song not given in couplets.

IN BRIEF

Behavior: Skulks in thickets.

ID: Gray with black cap and rust undertail. Length 8½ inches.

Habitat: Forest understory, thickets, gardens, backyards.

Nest: A bulky cup of sticks, grasses, leaves, grapevine shreds. The center of the cup is lined with rootlets, fibers, and plant down.

Food: About half animal matter — insects and their larvae — and half plant material — berries and seeds. At feeder takes orange halves and raisins.

carolinensis

GRAY CATBIRD   131

Toxostoma rufum

Brown Thrasher

Toxostoma

AS THE FRESH WHORLS of dogwood petals unfold along the woodland edges and riverine forest trails, the Brown Thrasher's distinctive song can be heard over two-thirds of the country. Though the notes have a harsh, raspy quality to them, the diagnostic feature is that they are sung in couplets or triplets: *"twee–twee toyou–toyou chack–chack wheep–wheep."*

Though a member of the mimic thrushes, it seldom imitates others and when an attempt is made it is poor at best. To find the singer, look for a perch high atop the tree and usually out in the open portion. It is a lovely bird, a rich foxy chestnut in color with heavy chestnut breast streaks. With head thrown back, its whole body seems to reverberate as it sings.

To return to the nest site, it will plunge from the treetops down into the dense underbrush below. In the thickest of thickets it will place its flat cup-like nest of sticks interwoven with grapevine bark. In western areas the nest may even be placed on the ground and it is often subject to heavy predation.

ATTRACTING THEM TO YOUR GARDEN

Thrashers will come to garden areas but are quite shy. In northern areas during the winter when an occasional bird will stay behind, suet is often a favorite item to help them survive until spring. Provide shrub cover and make water available.

DISTRIBUTION

Eastern two-thirds of the U.S. and south Canada.

THE SONG

twee-twee toyou-toyou chack-chack

wheep-wheep

- **Note sequence:** Always delivered in couplets which alternate between harsh notes and sweet whistles.

- **Time of song:** Morning and evening.

- **Other birds with similar song:** Similar to Catbird (*Dumetella carolinensis*), see page 130, and Mockingbird *(Mimus polyglottos),* see page 128, but delivered in couplets.

IN BRIEF

- **Behavior:** Sings from the top of a tree; forages and scratches about in thickets.

- **ID:** Foxy chestnut in color. Resembles a wood thrush but is larger and has a long tail, double white wing bars, and a long, curved bill. Length 11½ inches.

- **Habitat:** Brushy edges, thickets, ravines, woodland tangles.

- **Nest:** Large flattened cup of twigs, leaves, and bark tucked into a dense shrub or brier tangle.

- **Food:** Insects, other small invertebrates, some fruits and seeds.

rufum

BROWN THRASHER

Chamaea fasciata

Wrentit

Chamaea

THIS BIRD LIVES IN CHAPARRAL, a thick-leaved, resilient Mediterranean-type scrub that clothes the hillsides and coastal bluffs of central and southern California. As the name implies, it looks like a combination of a wren and a titmouse. It acts much like a titmouse as it works its way through the scrub in small groups, but its brown coloration and long cocked tail are reminiscent of a wren. In the southern portion of its range it is grayish, whereas in the northern area, it is a warm brown color.

The song is distinctive, a loud staccato "*pip-pip-pip*," then a descending, trailing-off "*ter-ter-ter-r-r-r*." The female also sings, but just the initial "*pip-pip-pip*."

Because of their secretive ways within the dense brush, these birds are invariably heard before they are seen. Occasionally the northern form will appear in an area totally out of the norm, such as foraging high in an oak. If the bird is hard to see, the nest is all but impossible to locate, tucked into the base of a shrub deep in the thickets. The eggs are as might be expected of a thrush — a lovely blue.

ATTRACTING THEM TO YOUR GARDEN
This shy species will occur in yards where chaparral is present, as they need dense cover. Water and fruiting shrubs may also attract them.

DISTRIBUTION

Central and coastal California, north to coastal Oregon.

THE SONG

pip pip pip ter-ter-terrrrrrr

- **Note sequence:** Often likened to a ball bouncing downstairs. Starts with several clear, crisp notes followed by a rapid succession of down spiralling slurred notes.

- **Time of song:** All day.

- **Other birds with similar song:** Song has been compared to the Canyon Wren's *(Catherpes mexicanus)*, see page 92, but is not as musical.

IN BRIEF

- **Behavior:** Skulks; moves about in loose bands.

- **ID:** Lightly streaked tan breast, creamy eye, and a long, unbarred and usually cocked tail. Short bill. Length 6½ inches.

- **Habitat:** Chaparral, coastal scrub.

- **Nest:** Well-made cup of plant fibers at base of brush clump.

- **Food:** Insects, especially hymenopterans, with fleshy fruits making up over half the diet outside the breeding season.

fasciata

WRENTIT   135

Anthus spinoletta

Water Pipit

Anthus

To MANY BIRDERS THE Water Pipit is a bird seen shuffling along a sandy beach, among the clods of a plowed field, or flying overhead in a roller-coaster flight pattern calling its name "*pi-pi-pipit*." To get to know this bird on its breeding grounds, one must travel north or head up into the mountains, where it is a common nesting species. Within this tundra setting the flight song, made up of a series of high pitched "*cheedle-zee*" notes, is reminiscent of a jingle of keys. Lacking any singing posts of height, the pipit has taken to the air for its place of territorial declaration. It circles high above often with a jerky flight pattern swinging in wide circles. Nearing the completion of the song period, which may last minutes, the bird plummets downward and lands on the tundra vegetation or rocky outcrop.

The nesting range is quite unusual. In the West, it is a bird of the High Rockies on up through Alaska. In the East, a small nesting population exists on the flat tableland of Mount Katahdin in Maine, representing the only nesting site in the eastern United States. In the fall and winter, flocks of thousands of birds can be seen in open fields across the southern portion of the United States.

ATTRACTING THEM TO YOUR GARDEN

Unlikely to visit a garden, but expansive lawns near agricultural fields could attract them. Provide water.

DISTRIBUTION

Breeds in high Arctic and mountains of Alaska and the Rockies.

THE SONG

pip-pip-pipit cheedle zee

- **Note sequence:** High-pitched whistled notes followed by a thin, lisping series. When flushed delivers pip-pip-it call.

- **Time of song:** All day.

- **Other birds with similar song:** Hard to confuse it with any other bird except Sprague's Pipit, which has a musical, twittering quality.

IN BRIEF

- **Behavior:** Pumps tail up and down; flashes white in tail as it takes off; large fall and winter flocks.

- **ID:** Brown back with faint streaking, light tan eye patches and underparts with darker streaks on breast and flanks. It often bobs its tail when it walks. Length 7 inches.

- **Habitat:** Open country, tundra, mountains above the tree line, fields, coastal flats.

- **Nest:** Grasses and sedges tucked into a grass hummock or rock crevice.

- **Food:** Insects and small invertebrates such as beach fleas.

spinoletta

Cedar Waxwing

Bombycilla

ON CRISP FALL MORNINGS, one often hears the faint high-pitched trill of this species as great numbers pass overhead in loose flocks bound for the South. Breeding season has ended and now the wanderings begin. Nesting can take place anytime within a period from early summer to early fall depending on the range frequented.

With their sleek plumage, black mask, yellow-tipped tail, and slim topknot crest they are unmistakable. The red, waxy tips can be seen when the birds are close by.

Although sedentary during the nesting season, massive flocks can build in the fall and winter. It is at this time that berries on trees may be stripped in very short order, fermented berries leading to a drunken stupor.

The song is a high-pitched, soft, trilled whistle that, once learned, is not forgotten, and is often the giveaway that the birds are present in an area well before they are seen.

ATTRACTING THEM TO YOUR GARDEN

The fruits of cedar, mulberry, and pyracantha are favorites but a wide range of other shrub and vine fruits are taken. Old apples and cherries placed out can attract them.

DISTRIBUTION

Throughout south Canada and the U.S.

THE SONG

_ _ _ _ _

see-see-see-see

- **Note sequence:** A high-pitched trill on one plane.
- **Time of song:** All day.
- **Other birds with similar song:** The larger Bohemian Waxwing has a harsher buzzy call.

IN BRIEF

- **Behavior:** Wanders about in groups; often seen chasing insects over ponds and rivers in the fall.
- **ID:** Warm brown with a crest and a black mask running through the eye. Yellow-tipped tail. Length 7 inches.
- **Habitat:** Open woodland, brushy areas, secondary growth zones; it is a wanderer; parks and gardens when on the move.
- **Nest:** Cup of grasses, rootlets, and bark strips often covered with lichen in the north of its range. Nest placed in branched fork.
- **Food:** Mainly berries and seeds but will switch to insects when plentiful.

cedrorum

Lanius ludovicianus

Loggerhead Shrike

140 | **SHRIKES: *LANIIDAE***

Lanius

WALKING THE FENCE LINES of a local farm when I was a boy, I came upon a rather macabre scene. The barbs of the wire held several specimens of dead animals, including a shrew and several grasshoppers. Perched atop a bush several yards away was the perpetrator of this mayhem, a Loggerhead Shrike. The strong, hooked bill is used to snatch and snare the prey while barbs of wire or thorns of trees such as hawthorn are used as the swords of final destruction. The food is left in place as a larder. Due to this unique feeding method, the nesting territory is defended and winter territories are established to assure a food supply and storage.

The shrike's favored site is a perch atop a bush or snag where it can see over open fields and grasslands. When prey is sighted, it drops from the perch, flies very low over the ground with a buzzy wingbeat and swoops upward onto the next perch.

The song is a jumble of harsh squeaks and guttural warbles delivered from a perch near the nest site.

ATTRACTING THEM TO YOUR GARDEN

The species is disappearing over most of its range. In the Southwest they appear in yards where there are spiny shrubs and plenty of insects.

DISTRIBUTION

Southern, west central and western U.S.

THE SONG

kerr kerr tulu tulu tulu karr karr

taleek taleek

Note sequence: Pairs and triplets of harsh non-musical notes.

Time of song: Morning and evening.

Other birds with similar song: Almost Thrasher-like, but rougher and choppy.

IN BRIEF

Behavior: Sedentary, rapid flyer low to ground; hangs prey items on thorns or barbs.

ID: Gray with a short hooked beak, a black face mask, and fast-beating wings. Length 9 inches.

Habitat: Open areas of all types usually with shrubs, includes farmland, prairie, desert, fields, and parks.

Nest: A bulky flat structure made up of a mass of twigs, grasses, and plant stems set in dense, low bush.

Food: Large insects such as locusts, birds and their young, small mammals.

ludovicianus

Warbling Vireo

Vireo

THE WARBLING VIREO is a drab, uniform gray bird with no wing bars and a hint of an eyeline. The western form at least has a yellowish hue to its sides. It frequents a wide variety of habitats, from the shade trees of a woodland lane to parks, gardens, and cottonwoods along streams. It seems that if there are fairly large deciduous trees, this bird will show up.

The song is a series of long, warbling notes given in succession, often on the same pitch. The male sings throughout the day from sunup to sunset and often sings on the nest! This is of course in direct opposition to the principles of nest camouflage and secrecy and makes it one of the easiest of all birds' nests to find.

The Warbling Vireo is very curious, and by making squeaking notes on the back of your hand, it can be enticed to come quite close, peering out of the leaves before giving a harsh, raspy "*twee-twee*" inquisitive scold. The species seems to show distinct peak and decline years, often with long intervening stretches.

ATTRACTING THEM TO YOUR GARDEN

Large trees in your yard are an attractant for this species. A bird bath may also attract them.

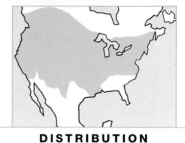

DISTRIBUTION

Throughout most of the U.S. and western Canada

THE SONG

twee-twee toosee toomee toosee torue

- **Note sequence:** Long series of melodic warbles dropping at the end.

- **Time of song:** All day.

- **Other birds with similar song:** The Yellow-throated Vireo's song is similar but more raspy.

IN BRIEF

- **Behavior:** Slow moving; vociferous songster.

- **ID:** Gray or gray-green back and white eye patches and breast. Length 5½ inches.

- **Habitat:** Large trees of upland woods, riverine forest, parks; tends to stay in upper branches.

- **Nest:** A pendulous cup of plant fibers and grasses hanging from crotch of branch sited moderately high up the tree.

- **Food:** Insects.

gilvus

Red-eyed Vireo

Vireo

ONE OF THE MOST COMMON SONGBIRDS of the upland woods is also one of the most constant singers. The Red-eyed Vireo is the world record holder for songs in one day with a reported 22,197 songs. The song is a rhythm of short phrases "*te-a-too tee-a-you*" that see-saw back and forth with a raspy quality.

As with all vireos, it moves about slowly in the canopy or sub-canopy vegetation with direct hops from limb to limb, scanning the foliage for insects and their larvae. A rather wide range in elevation may be chosen for the nest site. The nest is hung from the crotch of a shrub or tree limb from five to 20 feet off the ground.

This species winters in the Central and South American tropics, where it needs the protection of the tropical forest habitat. With the loss of this forest we have already seen a decline in the total number of breeding birds. If forest destruction continues in the U.S. and wintering grounds, what once was the most commonly seen songbird of the eastern forests may become a rarity within the next few decades.

ATTRACTING THEM TO YOUR GARDEN

Favors large flowering trees that attract insects. Provide water as most vireos bathe near dusk.

DISTRIBUTION

Across Canada, eastern U.S. and northern portion of western U.S.

THE SONG

too lee tee a too too la too see

- **Note sequence:** A series of two-parted whistles that rock back and forth.

- **Time of song:** All day.

- **Other birds with similar song:** Very similar to Blue-headed Vireo, which has a slower-paced song.

IN BRIEF

- **Behavior:** Deliberate yet slow moving; raises head feathers when disturbed.

- **ID:** Black line through the eye and edging the gray crown. White underparts. Length 6 inches.

- **Habitat:** Secondary growth woodlands with shrubby understory, also orchards, parks.

- **Nest:** A well-woven cup of plant fibers and grasses covered with lichen and bound with cobweb. Hangs from fork of outer shrub branches.

- **Food:** Mainly insects and small invertebrates.

olivaceus

Bell's Vireo

Vireo

BIRDING IN THE RIVERINE WILLOW thickets, bottomlands, or mesquite areas found in the central and southwestern states, you are sure to encounter the unique song of Bell's Vireo. Listening carefully, it sounds as if it is a question and answer series: the first part a distinct *"cheedle cheedle chee?"* with rising inflection, followed by an emphatic drop in pitch, *"cheedle cheedle chew."* The song is given over and over as the bird moves about the thickets or lower areas of the trees. When finally glimpsed, it is reminiscent of a Ruby-crowned Kinglet and is distinguished from the other vireo species by the less distinct wing bars and prominent eye-ring, which has a white line running to the bill from the eye.

The suspended, cup-like nest is placed in the fork of a branch and hangs down in the shape of an enlarged water droplet. This vireo suffers greatly from brood parasitism of Brown-headed Cowbirds, and the numbers appear to be on the decline in the western portion of its range.

ATTRACTING THEM TO YOUR GARDEN

Dense thickets with water available may attract them, certainly in migration. Flowering shrubs may attract insects for food source.

DISTRIBUTION

Central plains south through Texas and west through Arizona.

THE SONG

cheedle cheedle cheedle cheedle chew

- **Note sequence:** Sounds like a question and answer sequence.

- **Time of song:** Morning and evening.

- **Other birds with similar song:** Hutton's Vireo has a similar song sequence but is less rhythmic.

IN BRIEF

- **Behavior:** Active but shy; works all areas of trees and shrubs; fearless defender of its nest.

- **ID:** Those living in inland areas are brighter and yellower in color than their drab gray brothers living along the West Coast. Prominent eye-ring and white line from eye to bill. Length 4¾ inches.

- **Habitat:** Wet woods, especially riverine willow shrubberies and mesquite thickets.

- **Nest:** Cup of plant fibers covered with spider web in the branch of dense shrub.

- **Food:** Insects.

bellii

Oporornis formosus

Kentucky Warbler

Oporornis

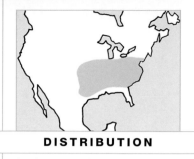

WATCHING A PARADE of tropical species approaching a birdbath at the edge of a Costa Rican rainforest, a familiar bird came from the woodland edge. Into the bath popped a beautiful specimen of a male Kentucky Warbler. The note, a low "*chuck*" with a hollow quality, is a fairly common sound in the thickets of the South. Here in Costa Rica, it was in basically the same type of habitat as its northern breeding grounds, where it also spends a large amount of time foraging on the ground. It overturns leaves and searches the base of shrubs for small insects, egg cases, and larval forms.

In the spring, the Kentucky Warbler returns to the North where its loud ringing song is heard once again. The song, although interpreted many ways, is basically a repeated, rolling "*tory-tory-tory-tory*" or "*turtle, turtle, turtle, turtle.*"

The nest is a bulky cup of leaves and grasses found on the ground, and often tucked under an overhang.

ATTRACTING THEM TO YOUR GARDEN

As a skulker, dense cover is attractive. Fruit trees such as a mulberry can attract them, especially in migration. Provide water for all warblers.

DISTRIBUTION

Southeastern U.S.

THE SONG

tory-tory – tory- tory

- **Note sequence:** Loud and rolling. "Hurry hurry hurry" or "tory-tory-tory".
- **Time of song:** Morning.
- **Other birds with similar song:** Similar to Carolina Wren found in same area, but lacking any raspy sections.

IN BRIEF

- **Behavior:** Shy and skulking, feeds on the ground most of the time, but usually sings in the trees.
- **ID:** Olive-green back and bright yellow underparts. Yellow eye-ring and line above the eye. Black mustache. Some of its crown feathers are tipped with gray. Length 5¼ inches.
- **Habitat:** Boggy woodlands, bushy swamps.
- **Nest:** A well-concealed, bulky cup of leaves and grasses on ground under overhang.
- **Food:** Insects and small invertebrates.

formosa

Northern Parula Warbler

Parula

THIS DIMINUTIVE WARBLER is a common species in migration and its diagnostic call makes it easy to locate. The call is a series of buzzy notes that climb the scale and finish with an abrupt "*zip*." The bird itself is usually found fairly high off the ground and near the outermost branches, where it hangs acrobatically like a chickadee foraging for food. It will also study leaves from below then flutter up to pick off a small caterpillar.

This scale of range can result in what appears to be a skewed distribution because nesting is often spotty. In the North, it prefers the coniferous forests where the amazing little cup nest is tucked within a dense Usnea clump. In the South, the draping masses of Spanish moss are to their liking, and the near-invisible nest is placed within the gray masses. Where neither of these growths are found the nesting choice varies, but usually an evergreen habitat is chosen.

ATTRACTING THEM TO YOUR GARDEN

In gardens of the South with large Spanish moss-covered trees, this species is sure to visit. Elm in the North and flowering oaks are also an attractant. During migration they may show up anywhere.

DISTRIBUTION

Canada to Florida and west to Texas.

THE SONG

see see see see se zip bzzzip

- **Note sequence:** Series of short buzzy notes that rise up the scale rapidly and end with bzzzip.

- **Time of song:** Morning.

- **Other birds with similar song:** The closest similar call is the Cerulean Warbler, which is more hurried at the end.

IN BRIEF

- **Behavior:** Fast moving; flutter feeds under leaves at medium and upper levels; gregarious.

- **ID:** Gray-blue back, yellow throat, narrow eye-ring and two white wing bars. The male has dark bands across its chest and a faint yellow patch on its back. Length 4½ inches.

- **Habitat:** Coniferous forests with Usnea lichen and mixed woods with Spanish moss; on migration, all forest types.

- **Nest:** Cup of fibers in Usnea or Spanish moss cup.

- **Food:** Insects, eggs, and larvae.

Yellow Warbler

Dendroica

THROUGHOUT MOST OF NORTH AMERICA where there is wet habitat in the form of a flood plain, willow bottomland, or moist tundra hillside, the loud and clear *"sweet-sweet-sweet-oh-so-sweet"* song of the Yellow Warbler is sure to be heard. In the spring when singing is at its peak, these birds can be seen in hot pursuit of aggressors crossing territorial lines or, with head thrown back in song, perched on the uppermost branches of a shrub. The males sing while collecting materials such as the fluff of willow seeds and bark fibers woven together to form the outside of the nest. The sturdy little cup is resilient enough to be seen through the winter following a nesting season.

The nests of the Yellow Warblers are favorite sites for the parasitic Brown-headed Cowbird to lay its eggs. However, unlike many other species, the Yellow Warbler on occasion will notice the undesirable egg and simply build a "floor" over it, increasing the nest's height, and re-lay.

ATTRACTING THEM TO YOUR GARDEN

Shrubs and flowering trees attract this common species. A bird bath is also an attractant. During nesting season, put out a mesh bag of short strings and thin yarn and the strands will be gathered and used for nest building.

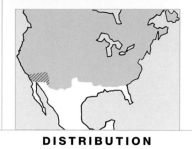

DISTRIBUTION

Across Canada and throughout the northern two-thirds of the U.S.

THE SONG

sweet sweet sweet Oh so sweet

- **Note sequence:** Very rapid and pure in quality.
- **Time of song:** All day.
- **Other birds with similar song:** In spring, the sweet song of the American Goldfinch *(Carduelis tristis)*, see page 228, is similar but longer in duration.

IN BRIEF

- **Behavior:** Very active in defending its territory; sings constantly, especially in the morning.
- **ID:** The male is brilliant yellow with a rust-colored beak, rust belly streaks, and black button eyes. The female is drabber and lacks the belly streaks. Length 5 inches.
- **Habitat:** Wet areas, especially with willows; open woodland, parkland.
- **Nest:** A fine-built cup of plant fibers, grasses, and down covered with willow threads and spider webs. Placed in crotch of a small tree.
- **Food:** Mainly insects, some spiders, and other small invertebrates.

petechia

YELLOW WARBLER

Chestnut-sided Warbler

WHILE WALKING THE FIELD EDGE or second growth area in the early spring, one will very often be greeted by a song that is best interpreted as a welcome: "*please, please, please to meet cha*." The greeter is the Chestnut-sided Warbler, a bird of the lower branches and tangle. It is a most striking bird in breeding plumage with a lemon-yellow cap and a black line through the eye and cheek area running into the rich chestnut brown of the sides. The underparts are pure white, the back streaked with light yellow. During the fall migration, it becomes a drab member of the migrating feeding flock; white below with greenish back and pure white underparts. There is often a smudge color to the sides. The eye is large and the cap has a very greenish color. As on the breeding grounds, it tends to forage low and frequent brushy thickets and slash areas.

With the coming of fall, the population wings its way south to Central and South America where it spends a lot of time quite high in the trees, often in the sub-canopy.

ATTRACTING THEM TO YOUR GARDEN

This is a bird of edges, so fields and open areas may be attractive. Bird baths will be used and flowering trees and shrubs may provide insect food.

DISTRIBUTION

Across southeastern Canada and north central and northeastern U.S.

THE SONG

please-please-please ta meet cha

Note sequence: A series of distinct phrases delivered in rapid sequence.

Time of song: All day.

Other birds with similar song: Somewhat similar to Yellow Warbler (*Dendroica petechia*), see page 152, but more distinct in phrasing.

IN BRIEF

Behavior: Very active and inquisitive. Usually feeds at low level.

ID: Chestnut-brown sides, yellow crown, black line through the eye, black mustache, heavily streaked back, and two yellow wing bars. Length 5 inches.

Habitat: Areas of secondary growth, shrubbery, thickets, overgrown meadows, fields, powerlines.

Nest: Loosely built cup of plant fibers and bark strips, low in bush or on ground.

Food: Insects, including defoliating caterpillars, spiders; in winter some seeds and berries.

Black-throated Blue Warbler

Dendroica

THE MASTERING OF BIRDSONG is a long, slow process. Therefore, it is most rewarding for the beginning birder to master a warbler song in very short order. Such is the song of the Black-throated Blue. It is very distinct in its uttering of "*I am so la—zeee!*" This trails on up the scale and ends with distinctive punctuation. To find the songster, look at fairly low understory or the lowest branches of trees in mixed woodlands, wet areas, and wooded, cool ravines.

The male is spectacular with its deep blue cap and back in sharp contrast to black face, throat, and sides and its pure white underbelly. Make note of the perfect square of white in the wing. This mark in itself is enough to identify this species in all plumages and for both sexes. The female is a dull brownish with buff-tinted underparts and the white primary spot stands out.

The nest is quite large for such a small bird. A fork in an evergreen shrub is most often chosen with the nest usually being placed within three feet of the ground.

ATTRACTING THEM TO YOUR GARDEN

Mainly seen in migration in evergreens and dense shrubbery. Will come to bird baths. Feeds on insects in flowering trees.

DISTRIBUTION

Eastern half of North America and south through the Appalachians.

THE SONG

I'm so lazy

- **Note sequence:** Of a buzzy, wheezy quality, distinct at first then rapidly rising up the scale.

- **Time of song:** Morning and midday.

- **Other birds with similar song:** Similar to an American Redstart (*Setophaga ruticilla*), see page 166, but slower and more emphatic at the end.

IN BRIEF

- **Behavior:** Forages at all levels but most common in the understory; often perches with wings partly open.

- **ID:** Male has blue back, black cheeks, throat, and sides; white underparts and small white wing patches. Female is brownish with the white wing patch. Length 5¼ inches.

- **Habitat:** Mixed woodland with understory.

- **Nest:** Bulky cup of leaves, sticks, and bark shreds.

- **Food:** Small insects, grubs and larvae, and spiders.

caerulescens

Pine Warbler

Dendroica

ON SCORCHING SUMMER DAYS in the dry, sandy pine barrens of the East, all is still under the blazing sun, all except for one bird whose solitary song dominates the atmosphere. The dry, staccato trill of the Pine Warbler is repeated over and over throughout the day, no matter how hot it gets. It is one of the least distinctively marked of our warblers, though the male has a rich yellow breast with faint side and flank streaks in spring.

It is a permanent resident in the pine flatlands of the South, and one of the earliest migrants in the northeast, its buzzy trill heard by the first week of April. As the name implies, it is a pinewood species, but more specifically it prefers areas with sparse growths of pitch pine and scrub oak. So, in order to be a prime habitat for this bird, pinewoods need to have a major fire every 10 years or so. In some areas, the problem has been solved by a policy of controlled burning to rejuvenate the woods, which has benefited the Pine Warbler greatly. It is still a relatively common and resilient species and is expanding its range.

ATTRACTING THEM TO YOUR GARDEN

Evergreens, in particular pines for nesting and foraging. Bird baths are used. Mealworms placed out in low containers have been taken.

DISTRIBUTION

Eastern half of North America.

THE SONG

— — — — — —

chee chee chee chee chee

- **Note sequence:** A continuous musical trill of chip notes.
- **Time of song:** All day.
- **Other birds with similar song:** Similar to a Chipping Sparrow (*Spizella passerina*), see page 208, but not as raspy.

IN BRIEF

- **Behavior:** Active high in the canopy; early arrival on spring feeding grounds.
- **ID:** Yellow breast, large white wing bars, white belly and undertail, and white spots on outer tail feathers. Length 5½ inches.
- **Habitat:** Pine woodlands, mixed woodland on migration.
- **Nest:** A cup of twigs, bark strips, and pine needles high in evergreen on a limb.
- **Food:** Insects, pine seeds, and some berries.

pinus

PINE WARBLER

Blackpoll
Warbler

Dendroica

THIS WARBLER is a true embodiment of the northern coniferous forests common throughout Canada and Alaska where its penetrating high lisping "*tsee–see–see–see–see*" call is sung throughout the day. Within the lower 48 states it nests only in the extreme northern section of New England, and northern New York.

It usually delivers its song from atop an evergreen, and its whole body seems to vibrate as it sits, head thrown back, delivering these penetrating notes. The spring birds are easy to identify with their bold black and white pattern and all-black cap or poll; the cheeks are white. Fall birds, however, can frustrate the warbler watcher faced with a complex of look-alike olive species. But note the wing bars, yellow tint to throat, white undertail, and orange feet.

In preparation for migration they fatten up to double their normal body weight. Out into the night, they pick up tailwinds (as far out as Bermuda) that will assist them on this final leg of their journey on a non-stop flight to South America.

ATTRACTING THEM TO YOUR GARDEN

Seen as a migrant. Feeds on insects in oaks and evergreens. Will come to bird baths.

DISTRIBUTION

Alaska across Canada to northern New England.

THE SONG

tsee – see-see-see see-see

- **Note sequence:** High-pitched and penetrating.
- **Time of song:** All day.
- **Other birds with similar song:** The Bay-breasted Warbler has a series of high-pitched notes but they are grouped in double series.

IN BRIEF

- **Behavior:** Stays fairly high in the trees, works its way out along limbs.
- **ID:** Black cap and white throat and cheeks. White underparts heavily streaked with black and white undertail. Two prominent white wing bars. Length 5½ inches.
- **Habitat:** Coniferous forests during breeding season, mixed forests during migration.
- **Nest:** Cup of twigs, rootlets, and lichen placed fairly high in the limb of a conifer on flat portion of limb.
- **Food:** Various small, winged insects, larval forms and egg cases, spiders.

striata

BLACKPOLL WARBLER

Mniotilta varia

Black-and-White Warbler

Mniotilta

IN THE SPRING, the Black-and-white Warbler is easily seen, as it is a bark gleaner and spends all of its time working on the trunk or limbs of the trees, often in clear view. In addition, its work takes it down to low levels in the understory trees for virtually eye-to-eye views. The Black-and-white remains in its "zebra" plumage throughout the year, though it is less intense in the fall, and is therefore a warbler that even the most novice birders can identify with confidence.

Its feeding mode makes it the "nuthatch" of the warbler group and this allows it to obtain food in a place that lacks competition from other warblers. In fact, it often moves with groups of warblers and chickadees during these foraging forays.

The song is fairly easy to remember. A drawn-out series of two-part, wheezy notes *"wee-see wee-see wee-see"* is likened to the sound of a small car starting. Occasionally it will break into a more complicated song, especially on territory. The nest is placed directly on the ground, usually in a concavity next to a rock or tree base.

ATTRACTING THEM TO YOUR GARDEN

Feeds in large trees, especially oaks and evergreens. A bird bath will be used for bathing and drinking.

DISTRIBUTION

Eastern half of North America.

THE SONG

wee-see wee-see wee-see

- **Note sequence:** A pulsating series of high lisping notes with greatest inflection on the first part of song.

- **Time of song:** Morning and night.

- **Other birds with similar song:** Similar to a Common Yellowthroat (*Geothlypis trichas*), see page 176, but much higher pitched and reedy.

IN BRIEF

- **Behavior:** Creeps about limbs of trees; often moves with mixed bird flocks.

- **ID:** The only warbler with black and white plumage and a white stripe through its crown. Length 5¼ inches.

- **Habitat:** Open deciduous and mixed woodlands, swamplands. On migration turns up in parks and gardens.

- **Nest:** Of leaves and grasses often with a canopy, usually placed at the base of a tree on the ground.

- **Food:** Insects, their eggs, larvae, pupae, and spiders gleaned from trunk and limb surfaces.

Wilson's Warbler

Wilsonia

PERPETUAL MOTION is one way to describe a
Wilson's Warbler. During migration, they dart here
and there, seeking small insects, hopping from limb
to limb, never seeming to sit still long enough for
the birder to get a good look. This is complicated
by the fact that they tend to stay low to the ground
and frequent dense thickets. When they do pop into
view, they certainly are one of the easiest warblers
to identify — bright yellow with an olive back and
a jet-black cap. They like to flutter feed, darting up
beneath a leaf with wings beating to snap up an
insect, and then dropping back.

Their song is an unstructured series of slurred notes
dropping rapidly in pitch, almost as if they wanted
to hurry up and finish the song. The nesting range
extends up into Alaska where Wilson's Warbler
is abundant, nesting in the endless alder thickets.
The nest is difficult to locate, usually tucked into
a hillside crevice, or at the base of a shrub such
as dwarf willows.

ATTRACTING THEM TO YOUR GARDEN

Low, dense shrubs and flowering trees may attract
this species in migration. Running water can also
serve as an attractant.

DISTRIBUTION

Across Canada and from Alaska
south in the Rockies and Cascades.

THE SONG

chee chee chee chet chet

- **Note sequence:** A wispy, high-pitched and rapid series, dropping off rapidly at the end of sequence.

- **Time of song:** All day.

- **Other birds with similar song:** Similar to Canada Warbler but more hurried and less complex.

IN BRIEF

- **Behavior:** Very active.

- **ID:** Olive-green back and yellow underparts with a dark eye on an otherwise plain face. The female lacks tail spots. Length 4¾ inches.

- **Habitat:** Thickets and scrub on migration, any area with dense cover.

- **Nest:** Small cup of grasses and fibers at or near ground level.

- **Food:** Small insects.

pusilla

WILSON'S WARBLER   165

American Redstart

Setophaga

THE REDSTART flits around among the foliage and dances along the limb, switching its body back and forth with each hop. Tail fanning in and out or flutter feeding under a leaf, it pursues its prey to the ground in an erratic flight, like a leaf tumbling to the ground. Bedecked in black and orange, the male is a handsome warbler. It is one of the earliest to arrive and one of the best-known because it is so common, and distinct yellow or orange squares in its tail make it easy to identify in all plumages.

The Redstart's nest seems to be a favorite of the parasitic Brown-headed Cowbird, and on several occasions I have seen this diminutive warbler stuffing larvae after larvae into the gaping mouth of its massive foster youngster.

In the fall they change into their protective colors — gray plumage, orange shoulders and flanks — and are one of the most abundant warblers on migration.

ATTRACTING THEM TO YOUR GARDEN
Favors hunting in oaks and flowering trees. Water availability is an attractant to all warblers. Dense shrubs and understory also favored.

DISTRIBUTION
From Alaska across Canada and the eastern half of the U.S.

THE SONG

zee zeee zee zee soowe

- **Note sequence:** A hurried buzzy song; a complex of lisping notes.
- **Time of song:** All day.
- **Other birds with similar song:** Similar to Chestnut-sided Warbler (*Dendroica pensylvanica*), see page 154, but without emphatic ending.

IN BRIEF
- **Behavior:** Very active, flutter feeds or works its way along a limb, tail fanned, swinging from side to side.
- **ID:** Mature males are jet black with orange markings that show brilliantly when they fan their tails. Length 5 inches.
- **Habitat:** Deciduous woodland with scrub and thicket understory, including parks and gardens with shrubs and young tree growth.
- **Nest:** A well-woven elongated cup of plant fibers and rootlets bound by spider web in crotch of scrub.
- **Food:** Insects, often small flying forms and spiders. Some fruit and seeds.

ruticilla

AMERICAN REDSTART

Protonotaria citrea

Prothonotary
Warbler

Protonotaria

A PROTHONOTARY is a keeper of records in the College of Prothonotaries Apostolic of the Roman Catholic Church. His robe color is a brilliant orange-yellow and one that the plumage of this warbler was compared with when first identified. Their range extends from Florida north as far as the Great Lakes, but in many people's minds this bird represents Spanish moss-draped stumps of the southern Bayou country and the low, flooded swamplands of the South. It is quite a wanderer and has appeared virtually coast to coast at one time or another during migration periods.

It has the uncharacteristic warbler habit of nesting in a natural tree cavity or abandoned woodpecker hole. The site chosen is fairly close to the ground, usually next to a backwater pool or sluggish stream. In this setting the loud and repetitive "*sweet-weet-weet-weet*" is staccato and on an even pitch. As with many other warblers, during the peak of the breeding season this bird will burst into the air and flutter about singing a canary-like song, hovering with feet dangling, usually near the nest site.

ATTRACTING THEM TO YOUR GARDEN

This shy species may appear in dense undergrowth and feeds on early mulberries in the South. Water is important. Spring wanderers in the North have appeared at feeders and taken suet and seeds.

DISTRIBUTION

Central and southeastern U.S.

THE SONG

sweet-weet-weet-weet-weet

- **Note sequence:** A sweet series of notes maintained at the same pitch.
- **Time of song:** Morning.
- **Other birds with similar song:** Somewhat like a Yellow Warbler *(Dendroica petechia),* see page 152, but not as jumbled at the end.

IN BRIEF

- **Behavior:** Stays low and near water.
- **ID:** Golden-yellow head and underparts, white undertail, blue-gray wings, and a blue-gray tail with large white patches. Length 5½ inches.
- **Habitat:** Swamps, wet lowlands, bayous.
- **Nest:** In natural cavity or old woodpecker hole.
- **Food:** Small insects.

citrea

Ovenbird

Seiurus

ONE OF THE MOST FAMILIAR woodland sounds for two-thirds of the United States is the loud sharp *"teacher–teacher–teacher–teacher"* rising up the scale as it intensifies. The calls may vary from an emphasis on the first note to the last. But in all cases the distinct *teacher* sound is hard to mistake.

The bird itself is more thrush-like than warbler-like when first seen walking about on the forest floor. Though they do sing while walking about, they seem to prefer to find a branch not too high above the leaf-strewn floor, to arch their head back and pour forth their song.

The name "ovenbird" comes from the nest, reminiscent of a Dutch oven and well hidden among the leaves on the ground. More often than not, it is located accidentally as you walk in the woods and are startled by a small bird fluttering up at your feet. The entry to the nest is a mere slit, making the whole structure seem like a bump in the leaves. Being a ground nest, many eggs and young are lost to ground predators.

ATTRACTING THEM TO YOUR GARDEN

A bird of the forest floor, they may stop in migration if dense shubbery is provided as well as fruiting shrubs and a water source.

DISTRIBUTION

Most of southern Canada, south through the midwest to the Carolina coast.

THE SONG

teacher-teacher-teacher-teacher

● **Note sequence:** A loud series of distinct phrases rising in intensity. Also a more gushing evening song.

● **Time of song:** Morning and evening.

● **Other birds with similar song:** Pulsing song is not to be confused with any others.

IN BRIEF

● **Behavior:** Forages about forest floor; will call from understory.

● **ID:** Olive-brown back, dull orange crown bordered by black stripes, white underparts streaked with black, pink legs, and clear white eye-ring. Length 6 inches.

● **Habitat:** Mixed woodland.

● **Nest:** Domed nest of grasses with side entrance.

● **Food:** Insects and other invertebrates.

aurocapillus

OVENBIRD | 171

Louisiana Waterthrush

Seiurus

IN EARLY SPRING, by the streams of ravine hillsides, the rich, full song of the Louisiana Waterthrush indicates that spring has indeed returned. It is one of the earliest of the warblers returning from its Central American and Mexican haunts where it has overwintered. The song is introduced by three clear whistles and followed by a series of sweet jumbled notes that cascade down the scale. The ringing quality of the song often makes it hard to pinpoint the singer, as it is delivered from a low tree limb, or from the rocks along the water's edge.

Foraging for insects and other small invertebrates, the bird diligently walks by a stream edge flicking over leaves and walking out into the water to turn over exposed debris. It looks very thrush-like — but sleek — and teeters its tail up and down as it walks. If you are very fortunate, you may see the bird make its way to the nest, tucked into an overhang of bank, or placed among the tangled roots of an upturned stump by the water's edge.

ATTRACTING THEM TO YOUR GARDEN

If you are lucky to have extensive gardens with rushing water, one may stop in migration. Wet edges near dense thickets would provide shelter.

DISTRIBUTION

Eastern half of the U.S. from New England to north Florida.

THE SONG

see-see-see-too la see de

- **Note sequence:** Two-parted, introduced by three clear whistled notes followed by a rapid series of clear bubbling notes rolling down the scale.

- **Time of song:** Morning and evening.

- **Other birds with similar song:** Similar to a Northern Waterthrush but not as rapid and with more introductory notes.

IN BRIEF

- **Behavior:** Bobs tail as it walks about; feeds at stream edge.

- **ID:** Brown back and heavily streaked underparts. Distinct white line over the eye and pure white throat distinguishes this species from its look-alike cousin, the Northern Waterthrush. Length 6 inches.

- **Habitat:** Breeds near flowing streams; other wet areas on migration.

- **Nest:** Bulky cup of grasses, leaves, and rootlets tucked into bank or upturned stump.

- **Food:** Invertebrates such as worms, as well as insects and their larvae.

motacilla

LOUISIANA WATERTHRUSH

Oporornis agilis

Connecticut Warbler

Oporornis

THE CONNECTICUT WARBLER was named by
Alexander Wilson who first saw it in Connecticut,
yet to this day it remains a scarce migrant to that
state during the fall. As it passes through, skulking
in the pepperbush and touch-me-not thickets of the
wetlands, it often goes unnoticed. In the spring, it is
a migrant west of the Appalachian Mountains and is
practically unknown on the Eastern seaboard during
this period. It resides almost entirely in Canada.
Outside of the western Great Lake states, it breeds
nowhere else in the United States.

It is a bird of the bog floor and walks about, much
like an Ovenbird, in search of small insects and
spiders. Its presence on territory is announced by
a loud, clear, ringing song that starts with two chips
moving on to a rapid "see-to-it see-to-it." The distinct
gray hood that crosses the chest coupled with the
complete eye-ring identifies this large warbler.
To really get to know this bird, expect to spend long
hours afield in wet habitat.

ATTRACTING THEM TO YOUR GARDEN

A shy and secretive bird that needs wet areas.
If your yard has moist ground where touch-me-not
and sweet pepperbush can grow, you may see one
in migration.

DISTRIBUTION

Southern, central Canada to north
Great Lakes region.

THE SONG

see to it see to it

- **Note sequence:** A loud ringing
  series rising in volume.

- **Time of song:** Morning.

- **Other birds with similar song:**
  Similar to Ovenbird *(Seiurus
  aurocapillus),* see page 170, which
  does not occur in the same
  breeding area.

IN BRIEF

- **Behavior:** Shy, solitary; walks while
  feeding; stays low.

- **ID:** The male has a gray hood while
  the female's is brown. Bold white
  eye-ring. Length 5¾ inches.

- **Habitat:** Cold northern tamarack
  swamps.

- **Nest:** Cup of grasses and rootlets
  on ground tucked among moss.

- **Food:** Insects.

Common Yellowthroat

Geothlypis

THE LOUD PLEASANT SONG *"twitchity, twitchity, twitchity"* of the Common Yellowthroat is a familiar sound of wetlands and coastal marshes throughout the United States and southern Canada. In action the bird is reminiscent more of a wren than a warbler. As it works its way about in the dense underbrush grasping the twigs from the side and peering out through protected openings, it even holds its tail cocked at an upward angle. During these times it usually scolds constantly with a sharp *"chick–chick."* The male is unmistakable with black mask, yellow throat, and olive upper parts. The female lacks the mask but has an all-yellow throat, marking the separation between it and look-alikes such as the mourning warbler. There are several races, but all have the same basic color pattern.

During the mating season, the male often bursts into the air with a fluttering flight, cocks its body upward and soars about giving a rather jumbled song of mismatched notes before plunging back into the thickets. This "ecstasy song" is performed by other warblers and seems to be a reinforcement of territorial behavior.

ATTRACTING THEM TO YOUR GARDEN

Will come to dense shrubbery bordering wet areas. Looks for insects in low flowering shrubs and will come to bird baths during migration.

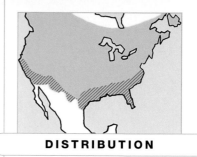

DISTRIBUTION

United States and southern Canada.

THE SONG

twitchity twitchity twitchity

- **Note sequence:** Series of rapid explosive calls with an emphasis on the first part: twitch---ity.

- **Time of song:** All day.

- **Other birds with similar song:** The Yellow Warbler *(Dendroica petechia),* see page 152, has a rollicking song but not as emphatic.

IN BRIEF

- **Behavior:** Very active but skulks amongst vegetation; cocks tail.

- **ID:** One of the easiest birds to identify; the male has a black face mask and yellow throat. The female also has a yellow throat, but her face is olive-green. Length 5 inches.

- **Habitat:** Damp fields, dense thickets, cattail marshes, mangroves in South.

- **Nest:** Rather large and often cone-shaped of grasses, large leaves, bark strips, on or near the ground.

- **Food:** Insects.

trichas

COMMON YELLOWTHROAT

Yellow-breasted Chat

Icteria

Although found from coast to coast, many people are not familiar with our largest warbler. This is mainly because it is a very shy species, and tends to be solitary outside the breeding season.

At 7½ inches, the Yellow-breasted Chat looks more like a thrush than a warbler. It inhabits the densest tangles, from grape vines ensnarled in brushy thickets to a labyrinth of impenetrable greenbrier. When approached, they often view the intruder over the shoulder so as not to expose the brilliant underbelly. They love to sing all night — a series of loud harsh "*chack chack–ree–chip chip–twee–twee*" in an endless flow of harsh jumbled notes. They also call during the day using a variety of harsh notes, clucks, gurgles, and whistles.

The nest is usually placed in the middle of a brush thicket. It is a rather bulky stick structure loosely put together and flattening very quickly once the adult visits continually with food.

ATTRACTING THEM TO YOUR GARDEN

Dense vine tangles of grape and Virginia Creeper can be an attractant. Provide water. In northern areas some birds will stay back during the winter. It is then they may sneak to a backyard feeder to steal a morsel of suet for high energy.

DISTRIBUTION

Most of the U.S. except portions of the plains.

THE SONG

chack-chack ree chip chip twee twee

Note sequence: A jumbled series of hacking and whistled notes of no set pattern, often delivered at night.

Time of song: Day and night.

Other birds with similar song: Could be confused with a Brown Thrasher (*Toxostoma rufum*), see page 132, but Thrashers always in duplicate couplets.

IN BRIEF

Behavior: Rises in air and sings with wings drooped and tail and feet dangling; solitary and secretive.

ID: Large bird with a long tail and a thick bill. Gray-green back, wings and tail, bright yellow underparts except for a white belly and undertail. Length 7½ inches.

Habitat: Thickets, wood edges, powerlines.

Nest: Bulky cup of twigs in thickets and vine tangles.

Food: Insects, spiders.

virens

Summer Tanager

"PEE–TUCK–I–TUCK" is a common sound in southern mixed oak and pine woodland. Once this call note is learned, the birder becomes aware of how widespread this species can be. Hopping into view, the male stands out like a glowing red light against the lush green foliage. With slow and deliberate movements, they work the lower canopy of trees for insects, stopping, peering, and then grabbing at a morsel.

This species really has a two-parted lifestyle. In the East they tend to be smaller and like oak-pine woodlands, whereas the western population prefer to live by water courses and cottonwood thickets, and tend to be larger.

This tanager normally winters in Mexico and Central America. Tanagers as a group are a neotropical species that, fortunately for us, have extended their range slowly northwards.

ATTRACTING THEM TO YOUR GARDEN

Though certainly not a feeder species, birds out of their normal range lingering in the North will often revert to feeders for food to help them make it through the winter. I have spent many days at a kitchen window in the Northeast watching a Summer Tanager feeding on suet, orange halves, and sunflower seeds while the snow of a December storm swirled in.

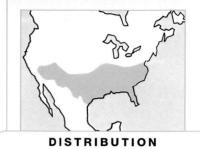

DISTRIBUTION

Southeastern and southwestern U.S.

THE SONG

pee-tuck-i-tuck/sear-to-wheer

to sear too whear

- **Note sequence:** Call is sharp, the song is harsh and raspy. Languid, see-saw phrases.

- **Time of song:** Morning and evening.

- **Other birds with similar song:** Similar to Scarlet Tanager (*Piranga olivacea*), see page 184, which is more raspy and choppy.

IN BRIEF

- **Behavior:** Fairly sedentary, stays in upper canopy; catches insects in the air.

- **ID:** The male has mottled green-red plumage during its first spring then turns red. Females are usually orange-yellow above and a dull yellow below, but some are a dull red color. Length 7½ inches.

- **Habitat:** Pine woodlands and mixed oak woodlands, swampy areas.

- **Nest:** Fairly flattened bulky twig nest often quite high in tree at end of limb.

- **Food:** Insects including bees and wasps.

rubra

SUMMER TANAGER

Piranga ludoviciana

Western Tanager

Piranga

As the name implies, this is a truly western bird. Its haunts are the open coniferous forests of both mountains and coastal lowlands. In such areas, as the birder wanders through the still forests, the strong call note is commonly heard — "*prid-it prid-it*," a clue that this sedentary bird is high above in the canopy. If the female is seen, she might be mistaken at first for a female oriole, but the heavier bill, darker upper parts, and pale cheeks make them easy to distinguish from each other. The nest will be placed very high in the tree on a flattened portion of limb and branch and is a simple cup of twigs.

Normally the species winters in Mexico and farther south. However, some birds stay as far north as Oregon, and others wander in an easterly direction and wind up on the East Coast.

In the woodlands, buckthorn berries prove to be the most favored plant during migration. At these times, willow thickets and oak bottomlands should be checked as the birds make their way south.

ATTRACTING THEM TO YOUR GARDEN

While in migration these birds do show up in gardens or at the feeding tray. Fruit is taken readily and cut-up orange halves with great delight. Raisins and dried cranberries have also been taken.

DISTRIBUTION

Western North America.

THE SONG

prid-it prid-it/tear tear cheer cheer

- **Note sequence:** The dry call note is most obvious; the song is strong, harsh, and coarse, in a two-part sequence.

- **Time of song:** Morning and evening.

- **Other birds with similar song:** Somewhat like Summer Tanager (*Piranga rubra*), see page 180.

IN BRIEF

- **Behavior:** Sedentary, moves about slowly in trees. Often comes to outer portion of limb, looks about and then takes flight.

- **ID:** Red head turns yellow-green in the fall, with yellow underparts, collar, and rump and black back, wings, and tail. The upper wing bar is usually a brighter yellow than the lower. Length 7¼ inches.

- **Habitat:** Coniferous forests.

- **Nest:** Bulky structure on flattened portion of limb.

- **Food:** Insects with berries later in the season.

ludoviciana

WESTERN TANAGER |

Piranga olivacea

Scarlet
Tanager

Piranga

THE BRILLIANT RED of the Scarlet Tanager's body sharply contrasts with the jet black of the wings and tail, making it one of the most striking North American birds to be found. The female is a lemon yellow with olive wings and has a black button of an eye. Tanagers can seem lethargic in their movements and are often overlooked as they move about feeding in the full-foliaged canopy. For this reason, they are thought of as much rarer than they actually are. One way to get a true feeling of how common they are is to learn their diagnostic song. It is a raspy, harsh "*hurry, to worry, flurry, its blurry.*" As diagnostic is the harsh call note "*chip–burr.*"

In the winter, the Scarlet Tanager heads for South America and early spring sees their return. Unfortunately their timing can be a bit premature, and at this time, spectacular build-ups of birds at the moderate coast habitats cover the ground and every tree. I have seen dozens huddled in the road or along the beach rack-line looking for insects. Some will perish but most will survive to again punctuate the spring woods with embers of glowing color.

ATTRACTING THEM TO YOUR GARDEN

As a visitor to the garden, it favors flowering trees (dogwoods/oaks/cherry) and will come to a bird bath. Not a feeder bird.

DISTRIBUTION

Northeastern North America.

THE SONG

hurry, to worry, flurry, its blurry/

chip-burr

- **Note sequence:** A series of harsh, coarse notes in rapid succession.

- **Time of song:** All day.

- **Other birds with similar song:** Like a Summer Tanager but harsher and more see-saw in tone.

IN BRIEF

- **Behavior:** Moves about slowly, feeds mainly in mid-canopy.

- **ID:** Male is bright red with black wing and tail. The female is olive-green above and green-yellow underneath. Length 7 inches.

- **Habitat:** Upland woods of oak, hickory, and maple, riverine forest.

- **Nest:** Flat, thin cup of grasses, roots, and twigs placed in a branched fork usually high up in the tree.

- **Food:** Mainly insects, especially caterpillars, also some berries and seeds.

olivacea

SCARLET TANAGER

Cardinalis sinuatus

Pyrrhuloxia

Cardinalis

THIS BIRD'S MOST UNUSUAL NAME, if broken down into its Greek and Latin parts, means "fiery-red bird with a crooked bill." What a striking bird it is, with brilliant red on its face, throat, and belly and a large, yellow, strongly curved parrot-like bill. Birders traveling to the Southwest for the first time often think that they are looking at a Cardinal on seeing the female. However, the grayness of the body, the yellow bill, and the exceptionally long crest with its bright red tip quickly eliminates the confusion.

This species has a limited range in the Southwest, and lives in a wide range of habitats from woodland edges to dry desert areas, mesquite shrub to garden thicket.

The nest, a fairly well-made cup of twigs, rootlets, and plant fibers, is tucked into a dense thicket. At the approach of an intruder, these birds become very alarmed and loudly scold man or animal with harsh "*chink*" notes and with their crest raised to a near-vertical position.

ATTRACTING THEM TO YOUR GARDEN

After breeding, they tend to form into small bands and move into agricultural areas to search out concentrated sources of seed. During such periods they will adopt suburban settings, and come in to the garden to feed. Like its cousin the Cardinal, sunflower seeds provide a main attraction.

DISTRIBUTION

Extreme southwest U.S. and western and southern Texas.

THE SONG

see-too see-too — pew you pew you

- **Note sequence:** A series of thin whistles and rapid chirps.
- **Time of song:** Morning and evening.
- **Other birds with similar song:** Northern Cardinal (*Cardinalis cardinalis*), see page 188, but thinner and not as rapid.

IN BRIEF

- **Behavior:** Stalks in thickets; but will pop into view to see intruder; will forage on open grass areas. Raises crest straight up when alarmed.
- **ID:** Gray back, red crest and red on face, throat, underparts, and tail. Thick, curved, pale-yellow bill. The female has a gray tail and more gray on its body than the male. Length 8¾ inches.
- **Habitat:** Weedy thickets, grassland edges, gardens.
- **Nest:** Compact cup of twigs, grasses, and plant fibers in a dense shrub.
- **Food:** Wide variety of seeds; will visit feeder for oil, sunflower seed, and cracked corn.

sinuatus

PYRRHULOXIA

Northern Cardinal

THE NORTHERN CARDINAL is a permanent resident throughout its range and nothing can quite compare with the brilliant red of this bird when seen against the newly fallen snow. It appears to be slowly inching west, so many more people will have the chance to enjoy this beautiful bird.

The male, in its vermillion colors with black mask and large bill set off with a distinct crest, is unmistakable. The female is also quite beautiful in her own right: warm brown buff with tints of red, and reddish in the wings and tail and at the tip of the crest. The bill is a bright berry red. The song is loud and clear and most distinctive: "*wheet, wheet, wheet*," loud, clear whistles with rising inflection, followed by a rapid series of "*purrty purrty purrty-chew chew chew*." The song can be delivered in any month but during the deep of winter is saved for sunny days.

ATTRACTING THEM TO YOUR GARDEN

The cardinal is a favorite of the backyard feeder; to increase your population, try to broadcast your seed more. Some birds have a narrow feeding range (many of the blackbirds, for example) others, like the cardinal, need more room. Therefore spreading sunflower seed in a limited area may attract only one pair. However, if scattered over a larger area, several pairs may come to feed. Remember though that cardinals are aggressive and can drive other birds away from a bird feeder.

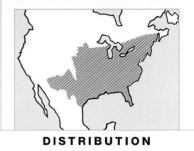

DISTRIBUTION

Eastern North America and from west Texas along the border through Arizona.

THE SONG

wheet wheet wheet purtty purtty purtty

— — —   —
chew chew chew

- **Note sequence:** A loud clear ringing series of whistles.

- **Time of song:** Morning and evening.

- **Other birds with similar song:** Individual whistled notes of Tufted Titmouse (*Baeolophus bicolor*), see page 82, can cause confusion.

IN BRIEF

- **Behavior:** Spends a good deal of time on the ground; perches on top of brush.

- **ID:** Brilliant red color with reddish conical bill. The female is less distinctive but can usually be identified by its bill which is similar to the male's. Length up to 9 inches.

- **Habitat:** Backyards, thickets, farmland, overgrown fields, woodland edge, parks.

- **Nest:** Loose cup of twigs, plant fibers, and grasses in dense shrub or tangles, not far off the ground.

- **Food:** The large bill is designed to crack open seeds and nuts; fruits, insects, and invertebrates also taken. Visits feeder readily for sunflower seeds.

Rose-breasted Grosbeak

As the pink of cherry and apple blossom tints the parkland and country lanes, the Rose-breasted Grosbeaks arrive. One of their favorite activities is to forage around these fruit trees and eat the ripening ovary of the flowers before they have a chance of setting into fruit.

The males are startling with a striking black hood and back running into the red glow of the breast patch and pure white underparts. The female, by contrast, looks more like a giant sparrow with warm browns, heavily streaked breast, and broad white and brown head marking. The bill is a massive cone designed for crushing fruit seeds. Even cherry pits are simple fare for such a massive structure.

In the mornings, the male will take to the treetops and deliver his territorial serenade — a series of rich, clear notes, rapidly delivered. The call note is a diagnostic *"chick"* given throughout the year and in the fall is often the only giveaway of their presence in the multicolored foliage. The female bird sings on occasion, a harsh, short version of the male's tune.

ATTRACTING THEM TO YOUR GARDEN

Attracted to fruit blossoms in spring, fruit and seeds in fall. Orange halves can also attract birds to visit your garden. Most leave the U.S. for Central America in the winter.

DISTRIBUTION

Western and south Canada south to midwest and mid-Atlantic states.

THE SONG

toodle loo tweedle leet ta do

Note sequence: Rapid sequence of clear whistled notes; in the fall a squeaking eek call.

Time of song: Morning and evening.

Other birds with similar song: Very similar to the Black-headed Grosbeak (*Pheucticus melanocephalus*), see page 192, but on a higher pitch.

IN BRIEF

Behavior: Sings from high in tree canopy, otherwise quite sluggish.

ID: A large bird, the male has a rose-colored throat patch and wing linings and black hood. Female is warm brown with streaked breast and white and brown stripes on the head. Length 8 inches.

Habitat: Open deciduous woodlands, parks, orchards, thickets.

Nest: Loosely made of twigs and grasses high in deciduous trees.

Food: Mixed insects and vegetable diet. Attracted to fruit blossom in spring, fruit and seeds in fall and winter.

Pheucticus melanocephalus

Black-headed Grosbeak

Pheucticus

IN THE SPRING when the Black-headed Grosbeak
returns from its Mexican and Central American
wintering grounds, woodlands, gardens, and
stream-sides of the West resound with the melodious
whistling song. This bird has become the sign of
spring in areas that show relatively little seasonal
change — when the grosbeaks are back, it is time
for other species to start up their spring serenades.

Male Black-headed Grosbeaks are magnificent with
their deep orange-cinnamon plumage, black-hooded
head, and wings of black flecked with white.
The female is more like a giant sparrow with her
bold white head streaks and a faint yellowish wash
to the underparts, which shows its close affinity
with the Rose-breasted Grosbeak. Hybridization
does take place in areas of overlapping range.
Both species also tend to wander to areas of each
other's range during aberrant migratory movements.
The males are easy enough to separate, but in
the females look for the yellow underwings of the
black-headed form compared to the red underwings
of the rose-breasted.

ATTRACTING THEM TO YOUR GARDEN

Provide shrubs and low fruiting trees. Bird baths
will be visited. Orange halves set out also attract.
Sunflower seeds are by far its favorite fare at a
garden feeder.

DISTRIBUTION

Western U.S. from British Columbia
south to Mexico and east to the
Dakotas, Nebraska, and west to
Texas.

THE SONG

to-we – to weer to weet-tee to wheer

Note sequence: A long series of
clear whistles.

Time of song: Morning.

Other birds with similar song:
Very similar to Rose-breasted
Grosbeak (*Pheucticus ludovicianus*),
see page 190, but lower in pitch.

IN BRIEF

Behavior: Slow moving, often
gives location away by giving off
a squeaky eeeak note.

ID: The male is distinguished by its
black head and orange-brown throat
and rump. Both sexes have a
yellowish belly. Length 7½ inches.

Habitat: Mixed woodlands, stream-
side edges, orchards, gardens.

Nest: Bulky stick nest in a tree fork.

Food: Various fruits and seeds; at
feeder prefers sunflower seeds.

melanocephalus

BLACK-HEADED GROSBEAK

Passerina amoena

Lazuli Bunting

WHILE THE EAST has its dramatically blue Indigo Bunting, the West has its counterpart, the Lazuli Bunting — an all-American bird bedecked in the nation's colors of red, white, and blue. Although it has a fairly wide range of habitats, brushy stream-sides and thick scrub chaparral seem to be the favorites. From a brush top, it delivers its rapid, jumbled song, not unlike that of the Indigo Bunting. In fact, in areas such as the Great Plains where the ranges of both the Indigo and Lazuli overlap, hybrids have occurred. This has led many to say that the species should be lumped together as one. Others say that without consistent hybridization, they remain separate species.

The female birds look remarkably similar, but the female Lazuli tends to have a bluish rump and the habit of flicking its tail, which aids identification. In earlier times, this was a popular cage bird, and in Mexico they are still captured while on their wintering grounds. Fortunately, new international laws governing migrant birds have put a halt to much of the songbird trading worldwide.

ATTRACTING THEM TO YOUR GARDEN

They are attracted to water and dense cover, so bird baths and flowering shrubs are important. A wide variety of seeds will be taken at feeders. Raisins will also be taken.

DISTRIBUTION

Western U.S. from Washington and Montana, south to southern California and northern Arizona and New Mexico.

THE SONG

seet seet – sue sue – seet seet

- **Note sequence:** A series of high-pitched paired notes.
- **Time of song:** All day.
- **Other birds with similar song:** Indigo Bunting (*Passerina cyanea*), see page 196, delivered at a faster pace.

IN BRIEF

- **Behavior:** Flicks tail constantly; sits atop a bush singing then slips into thicket.
- **ID:** Bright blue head and throat, reddish sides and breast and white belly. The plumage of the male resembles that of the Bluebird, except for the two wing bars and short, stubby bill. Length 5½ inches.
- **Habitat:** Brushy areas, chaparral, prefers water nearby.
- **Nest:** Cup of twigs, fibers, and bark strips in crotch of shrub.
- **Food:** Wide variety of seeds and fruits.

amoena LAZULI BUNTING

Passerina cyanea

Indigo Bunting

Passerina

THE DEEP INDIGO BLUE of this bunting is the result of feather structure and refraction of light, so there is no blue pigment in the feathers. Therefore, as the birds move about, the blue appears to twinkle as it goes from total indigo to black or gray. From their wintering ground in Mexico, the birds mass for their journey back to North America for nesting. By early May, they are dotted across the eastern half of the country in every overgrown field, orchard, or secondary growth land. They announce their presence by a loud song given in short bursts of paired high-pitched phrases. A constant singer, once on territory, they sing on through even the hottest hours of midday.

By late summer, the male begins to take on the mottled plumage of the winter adult: brownish in color with splotchings of blue. The female is a dull brown and lacks streaks. As they retreat south, thickets, garden plots, and weedy fields of agricultural land are scoured for seeds. As a great deal of land has been left to go fallow with the diminution of small-scale farm operations, the Indigo Bunting is profiting.

ATTRACTING THEM TO YOUR GARDEN

A wide variety of seeds are taken if they visit a feeder. Fruiting trees such as mulberry are favorites, and they will also come for orange halves.

DISTRIBUTION

Eastern half of U.S. with a population extending into southern New Mexico and Arizona.

THE SONG

swee swee-seet seet -

saya saya-seeo seeo

- **Note sequence:** A series of paired phrases delivered at a high pitch.

- **Time of song:** All day.

- **Other birds with similar song:** Lazuli Bunting (*Passerina amoena*), see page 194, but slower and more strident in tone.

IN BRIEF

- **Behavior:** Likes a high perch on treetop; gregarious on wintering grounds.

- **ID:** The male is deep blue but can look gray or black in poor light. The female is brown. Length 5½ inches.

- **Habitat:** Cut-over areas, re-growth burn sites, powerlines, weedy fields, orchards, parks.

- **Nest:** A shallow cup of twigs, leaves, and plant fibers placed in a shrub or dense brush tangle close to the ground.

- **Food:** A wide variety of seeds from grasses and other grains to composites.

cyanea

Spiza americana

Dickcissel

DICKCISSELS INHABIT RANK WEEDY MEADOWS and fields of the Midwest. The male is polygamous and therefore nesting is in loose colonies. The females build the small cup of grasses directly on the ground among grass and shrub bases or, rarely, up off the ground in the base of a shrub or on a grass clump. Large fields may have several such colonies.

The bird's name is derived from the call of the male, *"see see-Dick, Dickcissel, cissel."* In breeding plumage, the male looks like a tiny meadowlark. The females and immatures look like house sparrows with white throats and reddish shoulders. Although the majority of the birds winter from Mexico on through Central America, birds may show up at feeders on the East Coast. There they blend in with the house sparrow flocks feeding on seeds. With consistent searching of sparrow flocks, your diligence will be rewarded. On the winter grounds the birds mass by the thousands. Agricultural lands are a prime choice, as are stock-feeding yards.

ATTRACTING THEM TO YOUR GARDEN

A wide array of seeds are taken at the feeder. Bird baths are helpful to attract them for both drinking and bathing.

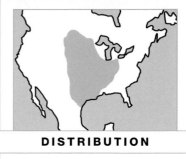

DISTRIBUTION

The Midwest.

THE SONG

dick dick cic cic cic sel sel

- **Note sequence:** A rhythmic unmusical series of wispy buzzy notes.

- **Time of song:** All day.

- **Other birds with similar song:** Similar to the Savannah Sparrow which lacks the distinct pulsing introduction.

IN BRIEF

- **Behavior:** Likes to sing from telephone and fence wires; very gregarious, masses in huge flocks in fall and winter.

- **ID:** Brown with rich rusty shoulders. Yellow on belly and face and around the eye and chest. Black throat mark. Length 6–7 inches.

- **Habitat:** Grasslands and weedy fields.

- **Nest:** Cup of grasses and rootlets usually at base of a grass clump.

- **Food:** Grass seeds, grains, insects. At feeder oil seed, sunflower, and fine cracked corn.

Pipilo chlorurus

Green-tailed Towhee

Pipilo

THIS HANDSOME SPECIES can be found in the dense brush and chaparral of the high plateau country of the West. Once located, the Green-tailed Towhee is secretive and will usually run quickly for cover and dart out of sight. However, inquisitiveness seems to overcome its initial fear, for the birds generally work their way back out into the open, often sitting in full view. The white throat is the giveaway, and looks as though a cotton ball has been tucked under the bird's chin. The top of the head is a rich rusty color, the back greenish-olive, and the underparts gray. As with all sparrows, the young birds are heavily streaked; the white chin, dark cap, and heavy back and belly stripes help identify the immature.

The Green-tailed Towhee has a call note much like the mew of a catbird. In addition, its double-foot scratching, typical of all towhees, can be heard as the bird forages in the underbrush. The song is a rich outpouring of notes with a harsh raspy portion in the middle of the sequence.

ATTRACTING THEM TO YOUR GARDEN

This is a shy bird, and prefers lots of shrubbery in which to hide. Planting flowering shrubs provides cover as well as nesting and perching sites, plus fruits and seeds to eat. In the winter there is a strong southerly movement but they are great wanderers from their normal migration route, and many easterners hope for their appearance at feeders in winter.

DISTRIBUTION

High plateau country of the West.

THE SONG

sweet-ooooo

- **Note sequence:** Opens with a clear, whistled sweet-oooo followed by a series of jumbled notes with a raspy central portion.

- **Time of song:** All day.

- **Other birds with similar song:** Gray Catbird (*Dumetella carolinensis*), see page 130, Fox Sparrow (*Passerella iliaca*), see page 214.

IN BRIEF

- **Behavior:** Secretive, often runs rather than flies into cover; scratches with both feet.

- **ID:** Green back, rust-colored cap, and white throat. Length 7 inches.

- **Habitat:** Dense shrubs of high plateau country.

- **Nest:** Large cup of grasses and bark shreds situated in dense scrub.

- **Food:** Wide variety of seeds, fruit, and some insects.

chlorurus

Eastern Towhee

THE EASTERN TOWHEE is a large sparrow with a long tail. The male has a distinctive black back and hood and has rich rufous chestnut sides. The underbelly is white. The female, who sings only on rare occasions, is brown on the head and back.

This is a common bird of low scrub and undergrowth from the plains to the forest edge of New England to the saw palmetto of Florida pinewoods. Recently, the Spotted Towhee of the west was designated as a fall species. Often while birding, the birds can be heard scratching in the leaf litter. If seen, they will be making a very quick move with both their feet at once raking out leaves with a double kick action.

The song of "*drink-your-tea*" once heard is not forgotten. Neither is the call that gives the bird its name — "*towheee*" — or one of its often-used vernacular names — "*chewink.*"

ATTRACTING THEM TO YOUR GARDEN

Dense cover, brush piles, and low shrubs are attractive to them. At the feeder they take sunflower seeds. Provide water for drinking and bathing.

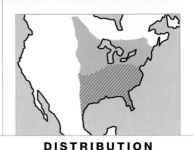

DISTRIBUTION

Eastern half of the U.S.

THE SONG

chewink/drink your tea

- **Note sequence:** A distinct call, and a loud clear song with the ending drawn out.

- **Time of song:** All day.

- **Other birds with similar song:** The Spotted Towhee — its western counterpart — is near identical.

IN BRIEF

- **Behavior:** Inquisitive; spreads tail as it hops about; scratches noisily with both feet among dead leaves.

- **ID:** Black back and hood, rufous chestnut sides, and white underbelly. Female has brown head and back. Length 8 inches.

- **Habitat:** Dry woods and riverside thickets, weedy hillsides, chaparral, parks.

- **Nest:** Leaves, bark strips, and twigs lined with fine grasses or pine needles, tucked into a depression on the ground under shrub cover.

- **Food:** Seeds and insects.

erythrophthalmus

Aimophila aestivalis

Bachman's Sparrow

On a cool spring morning in the dense palmetto stand that blankets the pinewoods floor, comes a sweet two-part song, reminiscent of the liquid notes of a Hermit Thrush. Scanning the dead branches that protrude above the palmetto, one encounters a rather drab-looking sparrow sitting with a vertical posture. It throws its head back and out pours the sweet cascade of notes. This is the pinewoods sparrow of old terminology in its perfect setting.

In the northern parts of its range, it favors abandoned fields and thicketed edges. Within this northern area, it seems to be slowly disappearing and retracting to its southern haunts. The reason for this is not totally clear. Getting a good glimpse of this bird is not always easy. It seems reluctant to sit in the open for any duration of time, and more often than not will pop up into view for a moment before twisting off in flight and diving back into the dense undercover.

ATTRACTING THEM TO YOUR GARDEN

It would be exceptional to have this species come to a feeder. However, in properties in the south with pinewoods and palmetto, the species may be resident and I have seen them come to bird baths placed on the ground at scrub edge.

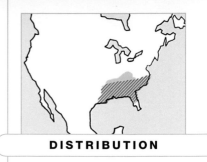

Southern U.S.

THE SONG

tooly tere ta see ta sa too lee

- **Note sequence:** A lovely rolling flute-like series of two-parted notes with rising, then falling inflection.

- **Time of song:** Morning and evening.

- **Other birds with similar song:** Hermit Thrush (*Catharus guttatus*), see page 118.

IN BRIEF

- **Behavior:** Shy, a skulker, hops up for one look and then is gone.

- **ID:** Similar to the Field Sparrow but with a darker bill and dark tail. Gray back streaked with dark brown, light gray underparts, and white belly. A western subspecies has dark chestnut brown plumage above and tan below. Length 6 inches.

- **Habitat:** Pinewoods and brush in North.

- **Nest:** Domed structure of grasses with side entrance at base of palmetto scrub.

- **Food:** Mainly small seeds.

aestivalis

BACHMAN'S SPARROW

Tree Sparrow

Spizella

For the birder who lives outside of Canada or Alaska, this bird is a winter visitor. In a good portion of the lower 48 states, it arrives with the cold nights of the October harvest moon. Weedy fields, brush piles, and marsh edges are usual first-selected sites, where the bird's sweet "*see-it*" or "*tweedle-it*" call gives its presence away. When it hops into view, it is quite easy to identify with its fox-red crown, two-tone black and yellow bill, and the distinct black "stickpin" spot on the clear gray breast.

The birds will linger on into the onset of spring, and by early April most are well along on their route back north. This deprives us of one of the finer types of songs given by the sparrow groups. Occasionally they will sing, and certainly on their northern breeding grounds it is one of the most common songs to be heard at the tree line. Introduced by clear "*sweet sweet sweet*" notes, a series of fluid warbling notes follows. The song is repeated over and over again. These birds nest north of the tree line and beyond out into the low thicketed valleys of the tundra.

ATTRACTING THEM TO YOUR GARDEN

As winter sets in, they usually move into more suburban sites and become common residents in and above the bird-feeding trays. Gathering up a wide variety of seed types, they do not seem particular in their food choices, although sunflower seed is a favorite for all the finches.

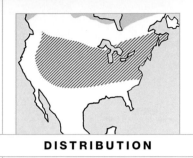

DISTRIBUTION

Alaska and northern Canada, winter across upper U.S.

THE SONG

see see – ta teedle ee tea tray

- **Note sequence:** A sweet-noted introduction, followed by a fast run of notes ending in a lower pitch.

- **Time of song:** All day.

- **Other birds with similar song:** Somewhat like liquid notes of Fox Sparrow (*Passerella iliaca*), see page 214.

IN BRIEF

- **Behavior:** Sings from top of trees and shrubs; in winter roosts in large flocks in marshes.

- **ID:** Largest of the rust-capped sparrows. Plain gray underparts, orange-chestnut back streaked with white and black, and black "stickpin" mark on chest. Length 6 inches.

- **Habitat:** In summer, scrubby tundra north of the tree line; in winter, weedy patches, marsh edges, and brush piles.

- **Nest:** Bulky cup of grasses and sedges lined with feathers and hair.

- **Food:** On breeding grounds mainly insects and seeds.

arborea

TREE SPARROW

Spizella passerina

Chipping Sparrow

Spizella

To the Easterner, the "Chippy" has become the familiar bird of the backyard, garden, or field edge. For the Westerner, it is a species that visits orchards but prefers life in the evergreen forests. Its trim form, brown back, gray underparts, chestnut cap, black line through the eye with distinct white line over it, and small black bill make this one of the easiest sparrows to identify.

Its song has given this bird its name: a long progression of dry rattle-like "*chipps*" given on one basic scale. Studies have shown that what to our ears sound like similar songs if slowed down via special tape-recording methods prove to be very distinctive in their delivery. Hence, when one stands in a cemetery (a favored habitat in the East) and listens to several of these birds singing from various evergreens, though the songs sound basically the same to us, each male can differentiate them and is thus well aware of the other birds' territory.

ATTRACTING THEM TO YOUR GARDEN

At feeders, they take a variety of seeds. Provide water. Also, put out mesh bags with combed-out doghair — they will use it to line their nests. Evergreens are favored.

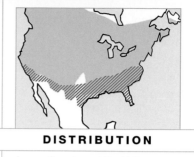

DISTRIBUTION

Across Canada and the U.S.

THE SONG

chip-chip-chip-chip-chip

- **Note sequence:** A staccato series of high-pitched chips, uttered more in early morning and afternoon.

- **Time of song:** All day.

- **Other birds with similar song:** Like a Pine Warbler (*Dendroica pinus*), see page 158, but more staccato.

IN BRIEF

- **Behavior:** Feeds a great deal on the ground; flocks in fall.

- **ID:** Distinguished by brick-red cap, light gray line over the eye, and black line running from the bill through the eye to the nape of the neck. Length 5½ inches.

- **Habitat:** Open areas of all types — lawns, parks, farmland, prairie.

- **Nest:** A tightly woven cup of grasses and rootlets lined with hair, usually on a branch of evergreen shrub or tree.

- **Food:** Small seeds of grasses and weeds, insects, spiders, and other invertebrates.

passerina

Calamospiza melanocorys

Lark Bunting

THIS IS THE SONGBIRD of sagebrush and dry prairie regions. A visit to such dry open areas is sure to reward the birder with views of the handsome males, all black in color with brilliant white wing patches, as they sail about on stiffened wings in courtship flight over their territory. As with many other birds of open regions, lack of perch availability for song means they have taken to the air to proclaim their springtime feelings. The song is a spirited outpouring of notes, although more commonly heard in migration is a distinct "*who-lee*" note.

The female is a streaky sparrow-like color with the distinctive white patch in its wing, although not as pronounced as in the male. Certainly one standout feature of this species is the very large bill. These birds feed on the ground and are fond of all types of seeds and many insects, especially grasshoppers.

In winter, massive flocks build up and often intersperse with plains longspur flocks in the South, especially Texas, where, in feed, hundreds and hundreds are not uncommon. In addition, a few representatives wander to the East and West Coasts each year.

ATTRACTING THEM TO YOUR GARDEN

On some occasions they may visit feeders. When they do, they prefer sunflower seeds but will take a mix of seeds as well.

DISTRIBUTION

Great Plains of the U.S. and south central Canada.

THE SONG

who-lee sweet sweet sweet

- **Note sequence:** A jumble of whistles, trills, and harsher notes.
- **Time of song:** Morning and evening.
- **Other birds with similar song:** Somewhat like the McCowen's Longspur, which nests in the same habitat.

IN BRIEF

- **Behavior:** Sings while hovering in the air; gathers in large flocks in the fall.
- **ID:** The male is black with a white patch on wings, white tips on tail feathers, and a thick gray bill. The female is streaked and has small tan wing patches. Length 7 inches.
- **Habitat:** Sagebrush, prairie, grasslands, agricultural land.
- **Nest:** Cup of grasses and rootlets on ground at base of grass clump.
- **Food:** Wide variety of seeds.

melanocorys

Ammodramus henslowii

Henslow's Sparrow

Ammodramus

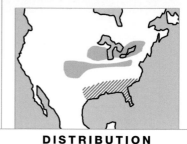

WHEN ONE GETS TO KNOW the Henslow's Sparrow, and that is no easy task, it is hard to believe that it is classed as a songbird at all. It is said to have the shortest song of any of the songbirds — a sharp, short, insect-like "*tis-lick.*" Though they do sing in the daytime, the low light intensity of evening initiates their main singing period. To add to the confusion, this is also the singing time for many crickets with a similar song. Both bird and insect will sing on and on during still spring and summer nights.

A resident of the shrubby, weedy meadows and fields of the northeast and central states, its choice of nesting areas is quite spotty. The populations may move from one site to another, use it for several years, and then disappear for a time before reusing it, or simply disappear for good.

These sparrows act more like mice than birds. In several instances, I have closed in on this species to have them escape by running between my legs or over my feet. If flushed, they fly for only a short distance before diving back into the grasses.

ATTRACTING THEM TO YOUR GARDEN

This would be a stunning bird to see at your feeder or in your garden, as it is very secretive. It would feed on seeds, including grass seeds.

DISTRIBUTION

Northeast and central states.

THE SONG

tis-lick

- **Note sequence:** A short, sharp double note with rising inflection.

- **Time of song:** From dusk through the night.

- **Other birds with similar song:** Unlike any other bird's song. More like a cricket's chirp.

IN BRIEF

- **Behavior:** A skulker — creeps about mouse-like in grasses, is reluctant to fly. Flies jerkily twisting its tail.

- **ID:** Streaked breast, olive nape and hind crown, and chestnut-brown wings. Heavily streaked back. Underparts are white with tan on the flanks and breast is streaked with black. Length 5 inches.

- **Habitat:** Wet scrubby fields, old meadows, salt marshes.

- **Nest:** A cup of grasses and plant fibers tucked well into the base of a grass clump.

- **Food:** Seeds and small insects.

henslowii

Passerella iliaca

Fox Sparrow

Passerella

For a good portion of the United States, this is a winter feeder bird. When it does arrive, its size and fox-like color certainly make it stand out against the white blanket of snow that drives it into the feeders from nearby thicket and weed fields. It is not only one of our largest sparrows, but one of the handsomest. There are many subspecies, however, and in coloration the bird can be bright rust or dark chocolate brown, as seen in the Pacific Northwest forms. The one feature that is consistent is the rusty tail. They scratch about in thickets with a double foot kick method, and on first appearance can lead one to think of a Hermit Thrush. However, the large bill immediately eliminates that species.

The birds can be quite shy and quickly slip from their shrub-top perch at the first sign of an intruder. The more open the area inhabited, the shyer the birds, so one must take quick advantage of any available sparse cover. The West Coast birds are often encountered as they scratch about the paths through the coastal brush and chaparral.

ATTRACTING THEM TO YOUR GARDEN

The large bill of the sparrow is designed for feeding on seeds and fruits, and much of their winter movement is based on the availability of these items. Will take sunflower seed at a feeding tray as well as other seed mixes. Provide shrub cover and water.

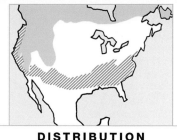

DISTRIBUTION

Northern Canada and Alaska, and western Canada south in moutains to Nevada and Colorado.

THE SONG

ta tee ah oo la tee tay ah

- **Note sequence:** Series of rich clear notes in single slurred arrangement.

- **Time of song:** Morning and evening.

- **Other birds with similar song:** Like a Tree Sparrow (*Spizella arborea*), see page 206, or a Goldfinch (*Carduelis tristis*), see page 228, but a clearer tone.

IN BRIEF

- **Behavior:** Secretive, gregarious outside breeding season; double scratches with both feet at the same time.

- **ID:** Varies widely in color and pattern, but most have heavy markings on the underparts, merging into a spot on the upper breast and are reddish-brown to gray-brown above. Rusty tail. Length 7 inches.

- **Habitat:** Conifer and mixed woods, willow thickets and tundra edge, chaparral in West.

- **Nest:** Large cup of grass, rootlets, and leaves lined with hair or feathers, set close to the ground.

- **Food:** While nesting, mainly insects. At other times seeds and fruits; oil seed and sunflower at feeding tray.

iliaca   FOX SPARROW

Pooecetes gramineus

Vesper Sparrow

IN EARLY TIMES known as bay-winged bunting and spectacled bunting, these two names point out two key features of the bird we know as the Vesper Sparrow. The rusty bay patch of the wing is often conspicuous, and the large white eye-ring certainly is. In addition, when flushed, the outer tail feathers flash white.

The term "vesper" was coined because those who heard its song felt that the evening was the best time to appreciate it. In fact, the species sings throughout the day, and happens to have one of the more charming songs of all the sparrows. It is a clear whistled series that begins with distinct phrases that then rapidly trail off. There are many interpretations, but the basic pattern is "*tear–tear, tore–tore–here–we–go–down–the–hill*." These songs are usually delivered from perches higher than the normal grass clump top, such as low tree branches at field edges or fence posts.

Its choice of habitats — sand plains, low grass areas, prairies, and beach-grasses — unfortunately represent waste areas ripe for use by land developers, which is why the Vesper Sparrow is disappearing over many places within its range, especially in the East.

ATTRACTING THEM TO YOUR GARDEN

Reluctant to come into gardens but during harsh winters may come to seed. Provide low bush cover.

DISTRIBUTION

Across northern North America.

THE SONG

tear-tear tore-tore-here-we-go-

down-the-hill

Note sequence: A progression of two phrase sequences that scale downward in rapid succession.

Time of song: Morning and evening.

Other birds with similar song: Similar to a Lincoln's Sparrow, which has a sweet, bubbly trill.

IN BRIEF

Behavior: Alert, often sings from a fence post or low tree limb at clearing edge.

ID: Distinguished from other sparrows by short notched tail with white outer feathers, chesnut-brown on wing and white eye-ring. Also has dark cheek patches and two inconspicuous wing bars. Length 6 inches.

Habitat: Open areas — grasslands, sand and gravel pits.

Nest: Of grasses and plant fibers on ground.

Food: Grass seeds, berries.

gramineus

Melospiza melodia

Song Sparrow

THE SONG SPARROW makes its home in thickets and dense scrub from coast to coast and throughout most of Canada. It is a bird of the backyard, the park, and the woodland edge. Fortunately, it has a lovely song and it is fairly easy to learn. Three or four clear introductory notes are followed by a rolling series of sweet notes ending in a buzzy trill.

It is a long-tailed sparrow, and when seen fleetingly as it disappears into a shrub or wood pile, the elongated tail seems to spiral into the air just as it slips out of sight. The species is highly variable. Some 30 subspecies are represented in North America, ranging from the large dark form of the Aleutian Islands of Alaska to the pale forms of the southwestern deserts.

Earlier nesting birds almost always place the cup of grasses, wet stems, bark, and rootlets directly on the ground beneath a sheltering shrub or low-growing plant. The birds nesting later in the season tend to be more low-tree and shrub nesters. At the nest, both parents tend the young and often the male is still on young-raising duty when the female is sitting on the second set of eggs for the season.

ATTRACTING THEM TO YOUR GARDEN

Low fruiting shrubs will provide food and cover. They enjoy a bird bath and will usually bathe in the morning and evening. They come to feeders and will take a wide variety of seeds, especially sunflower.

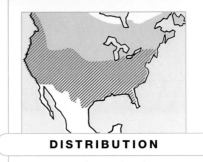

DISTRIBUTION
Across most of North America.

THE SONG

see see see – till a da see see

Note sequence: Highly variable. A series with sweet introductory notes followed by a rapid series cascading downwards and ending abruptly.

Time of song: Early morning and late afternoon.

Other birds with similar song: Similar to a Fox Sparrow (*Passerella iliaca*), see page 214, which also has sweet introductory notes followed by trills.

IN BRIEF

Behavior: Hops about in brush piles and weed fields; is attracted readily by "squeaking."

ID: Grayish face with rich chestnut head stripes, chestnut back and heavily streaked chest, and sides of dark brown. A concentration of two small and one large chest spots merge at a distance to give the identifying "stickpin" field mark. Length 5½–7 inches.

Habitat: Thickets, brushy areas, old fields, gardens.

Nest: Cup of grasses and rootlets with finer grass lining.

Food: Seeds and small insects.

melodia

SONG SPARROW   219

Zonotrichia albicollis

White-throated Sparrow

Zonotrichia

THE WHITE-THROATED SPARROW is a large, handsome sparrow with rich brown rust on the back and distinctive white and black head stripes. The throat is pure white on adults and dusky gray on the immatures. Note the area between the eye and the base of the bill called the "lores." In top plumage, this is a rich lemon yellow.

Their nesting areas lie mainly in Canada, the boglands of New England, and the Appalachians in the evergreen zone. It is on their breeding grounds that the sweet diagnostic song is best heard. It is a long, drawn-out but melodic "*oh sweet-sweet-poverty-poverty-poverty*." The song varies depending on what part of its range you are in. On migration and on its wintering grounds, it inhabits thickets and understory of woodlands where the sweet "*tseep*" is an indicator that a foraging flock is near.

This species is polymorphic — it has two color phases. In one, the head streaks are pure white, and in the other they are buff. Check to see which comes to your feeder next fall.

ATTRACTING THEM TO YOUR GARDEN

This is a classic feeding-station bird. They occur in large numbers, and will stay in or near a yard for an entire winter if the feeding sequence is on a set regime. At the feeder, they love sunflower seeds and peanut hearts.

DISTRIBUTION

Across Canada and eastern North America.

THE SONG

oh sweet sweet poverty poverty

- **Note sequence:** A series of clear loud ringing whistles.
- **Time of song:** Morning and evening.
- **Other birds with similar song:** Similar to the White-crowned Sparrow (*Zonotrichia leucophrys*), see page 222, which has a shorter, less distinctly phrased song.

IN BRIEF

- **Behavior:** Scratches noisily for seeds on the ground; gathers in small flocks in the fall.
- **ID:** Rich rust back with black and white striped head and pure white throat. Length 7 inches.
- **Habitat:** Edges of coniferous woods, in bogs and wet grassy areas, weed fields, brushy hillsides and swampy areas, also parks.
- **Nest:** Cup of fine grasses, plant fibers, and rootlets built at the base of a clump of grass or moss hummock.
- **Food:** Weed seeds, insects. At feeder loves oil seed and sunflower.

albicollis

WHITE–THROATED SPARROW

White-crowned Sparrow

Zonotrichia

WITH ITS BLACK CAP and white stripes jauntily tilted forward, large size, and grayness of body, an adult White-crowned Sparrow is a striking bird and one of the most common species of the West Coast and sub-Arctic North. One of the true heralds of spring in the open tundra, the swale is quickly alive with the sound of the clear introduction notes, followed by a rapid run of twittering down-slurred trills. The birds seem to be atop every small snag and continue to call all through the long sub-Arctic days.

Farther south along the West Coast, the same song is heard off and on throughout the year, but begins in earnest in the early spring. Here, the White-crowned Sparrow is also a common species of weedy thickets, brushy areas, parklands, and gardens.

For the Easterner, it is a bird of passage headed for more southerly wintering grounds. The majority of the migration south is through the Great Plains and mid-Atlantic states, so the bird remains uncommon in the Northeast. Sightings are predominantly of young birds with rusty crown stripes.

ATTRACTING THEM TO YOUR GARDEN

A maintained brush pile at a garden's edge can often attract a nesting pair of these most handsome sparrows, and they come readily to winter feeders for seeds, including millet.

DISTRIBUTION

Across northern Canada and south in the Cascade and Rocky Mountains.

THE SONG

seet see see – toodle ee seet

- **Note sequence:** Clear whistles followed by a rapid run of liquid notes.

- **Time of song:** All day.

- **Other birds with similar song:** Similar to the White-throated Sparrow (*Zonotrichia albicollis*), see page 220, but lacks the distinct phrasing of notes and is shorter.

IN BRIEF

- **Behavior:** Bold and active; sings from elevated perch; aggressive near nest site.

- **ID:** Gray body with black cap and white stripes. Length up to 7 inches.

- **Habitat:** A wide variety of scrub, and shrubby areas of secondary growth and weedy fields.

- **Nest:** Sheltered by grass and moss clump on the ground, made up of grasses and rootlets lined with hair.

- **Food:** Insects and berries, seeds in winter. Comes readily to winter feeder for seeds including millet.

leucophrys

WHITE-CROWNED SPARROW 223

Junco hyemalis

Dark-eyed Junco

For most of the Eastern Seaboard and the South, the Dark-eyed Junco arrives with the first cold blasts of northern winds in the fall. Among the colorful leaves of roadsides, small parties of these birds flush up into the trees as cars pass. The white edging to the tail is diagnostic enough to make the identification. In fields and woodland edges, they mingle with mixed groupings of sparrow relatives as they forage for seeds, often calling in concert with a varied warble-twitter, or odd "*clunk*" or "*smack*" sounds.

The species has gone through a remarkable number of name changes and shows dramatic color changes depending on the geographical location of the population. The West is dominated by the "Oregon form" with rusty back and contrasting black hood. There is the "gray-headed form" of Arizona and Mexico with its rusty back; and the isolated Black Hills of South Dakota are the habitat for the "white-winged form." In the East, the old name of "slate-colored" still holds sway.

ATTRACTING THEM TO YOUR GARDEN

With the first heavy snows, the Juncos wend their way to the backyard feeding stations. They are principally ground feeders and are usually quite reluctant to go up to a feeding tray, so seeds should be tossed in snow-cleared areas. They will feed on sunflower and other smaller seeds.

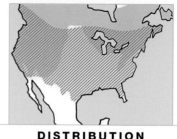

DISTRIBUTION

Across Canada and western U.S. as well as New England and the Appalachians.

THE SONG

tea tea tea tea tea tea/sweet sweet sweet

- **Note sequence:** Rapid series of notes delivered on the same pitch; call is a rapid twitter.

- **Time of song:** Morning and evening.

- **Other birds with similar song:** Somewhat like a Chipping Sparrow (*Spizella passerina*), see page 208, but not as rapid or harsh.

IN BRIEF

- **Behavior:** Active ground forager, gregarious; flashes white on tail edges when it flies up.

- **ID:** Once several species, this bird is now recognized as one species with different forms. All forms have pink bill, white outer tail feathers, white underparts, and dark eyes. Length 5½ inches.

- **Habitat:** Mixed woodlands, coniferous forests, wet areas, badlands; gardens in winter.

- **Nest:** A cup of grasses, rootlets, bark strips, and hair tucked into bank under rock overhang, in tree roots.

- **Food:** Mainly insects in summer and seeds in winter — millet, canary seed, fine cracked corn, and oil seed.

hyemalis

Carpodacus mexicanus

House Finch

THE HOUSE FINCH is the bubbling songster of the backyard and courtyard. The population has boomed and now reaches from coast to coast. Introduced in the early 1940s to the East Coast, the population has made a dramatic explosion in the last 25 years, and during the last part of the century the western population and the eastern population finally joined on the prairies. The eastern population rarely show the orange coloration of many southwestern regions.

This abundant species favors weedy areas, brushy hillsides, and arid zones in the West. Farmland and grassland edges are other favorite spots. The House Finch likes to live near humans and sometimes spends its entire life in one area, rewarding its neighbors with beautiful song. On warm mornings, even in winter, the bird's sweet rapid jumble of notes cascading from an ivy-covered wall is uplifting.

ATTRACTING THEM TO YOUR GARDEN

One way to attract nesters is to affix an old strawberry box under an overhang — a prized site for this backyard bird — and scatter nesting materials such as mesh bags of hair and bits of yarn. A selection of seeds, berries, and fruits will attract them to the feeder. Peanut butter with seeds is readily taken, as well as fine cracked corn, millet, peanut hearts, and thistle seeds. If you ensure a sufficient supply of water, all of their needs for backyard residence will be met.

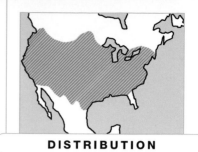

DISTRIBUTION

Coast to coast and north to south.

THE SONG

to da see to ta do see

lee lee to da to da to lee

- **Note sequence:** A jumbled series of sweet notes and whistles interspersed with slurred harsh *wheer* notes dropping in pitch.

- **Time of song:** All day.

- **Other birds with similar song:** Swainson's Thrush (*Catharus ustulatus*) and Purple Finch (*Carpodacus purpureus*), see pages 116 and 234.

IN BRIEF

- **Behavior:** Very vocal, sings for long periods.

- **ID:** Male has red plumage on forehead, breast, and rump; female is streaked brown; 6 inches long.

- **Habitat:** Frequents areas around human habitation, weedy areas, bushy hillsides, and wooded canyons.

- **Nest:** Bulky cup of grass, fibers, and bark strips placed in conifer or ornamental shrub.

- **Food:** Mainly seeds; sunflower seeds at feeder.

mexicanus

American Goldfinch

Carduelis

IN THE WINTER MONTHS, the American Goldfinches, cloaked in drab olives with black wings that sport white barring, cling to the feeders and hang in every which way as they split the seeds with their conical bills. As spring approaches, the yellow color slowly starts to show as plumages wear. (Wear plumage is not common in the bird world, but this species is a fine example of the process.) By spring, they are back in their brilliant golden garb and wandering about in loose flocks. Trees often come alive with the sweet twittering songs of numerous males. In all cases a distinctive "*swee*" note in a rising inflection can be heard identifying the songster. Though at their finest during the spring, nesting is often postponed until well into the summer because fine thistledown is a favorite addition to the newly completed nest. The nest is a solid, wool-like structure placed in the fork of a tree branch and made of plant fibers and cotton, and with spider and caterpillar nest webbing.

ATTRACTING THEM TO YOUR GARDEN

After nesting, the finches group together to look for feeding stations loaded with thistle seeds. In addition, planting sunflowers is a sure way to attract them. A garden plant that is also favored is the Cosmo. Once the planted sunflowers are used, they will come to feeders filled with sunflower seeds.

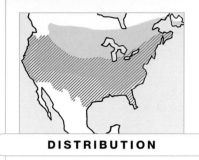

DISTRIBUTION

Across North America.

THE SONG

perch-a-week

- **Note sequence:** The flight song is perhaps the best known, a three-parted perch-a-week. Otherwise song is a jumble of sweet whistles and twitterings.

- **Time of song:** Morning and evening.

- **Other birds with similar song:** Close to Purple Finch (*Carpodacus purpureus*), see page 234, but not as harsh.

IN BRIEF

- **Behavior:** Lively, moves about in groups. Has roller-coaster type flight, utters flight note at top of rise.

- **ID:** The male is brilliant yellow with a black cap which turns brownish-gray in winter. Length 5 inches.

- **Habitat:** Open fields, weedy areas, roadsides, farmlands, gardens.

- **Nest:** A beautiful cup of grasses and plant fibers in the fork of a shrub, lined with thistledown.

- **Food:** Mainly seeds but insects taken during nesting.

tristis

Red Crossbill

IN THE LUSH FOREST of the Mexican mountains, I heard a familiar sound: Red Crossbills. They swung in and came to rest in the top of the pine. I could see their unique crossed mandibles, used for opening evergreen cones. Many would be surprised, believing the Red Crossbill to be a northern species. But looking at its full range, one can see that it lives throughout the high areas of the West all the way down through the Mexican mountains. Indeed, there is even a relic population in the mountains of Haiti!

Red Crossbills are very irregular in their wanderings and basically move as the cone crop varies. In eastern regions, several winters may pass before an influx occurs. When it does come, it often seems that the distinct "*jip–jip–jip*" notes of the roaming flocks can be heard everywhere. They are often fond of saltings and gravelling at roadsides. Arrivals from the North are often unbelievably tame, allowing one to approach up to a few inches. At banding operations, I have actually picked birds off the limbs of trees to band them.

ATTRACTING THEM TO YOUR GARDEN

At the feeding tray, sunflower seeds will disappear quickly. They will also come into gardens to eat the seeds of conifers and other fruits and seeds. If you do not have conifers, collect cones that still have seeds behind closed bracts and hang them near the feeder.

DISTRIBUTION

Across Canada and southwestern U.S. mountains to Mexico.

THE SONG

jip-jip-jip jip-jip-jip

Note sequence: The distinctive call consists of a series of sharp notes, while the song is a softer too-tee too-tee too-tee tee tee.

Time of song: Morning.

Other birds with similar song: Close to White-winged Crossbill, which has a harsh chit chit call.

IN BRIEF

Behavior: Gregarious, tame; will visit conifers in gardens.

ID: Distinctive crossed bill. Length 6½ inches.

Habitat: Coniferous woodlands.

Nest: Cup of twigs, rootlets, and grasses, lined with mosses, fur, and feathers.

Food: Seeds of conifers, also other fruits and seeds (maple, beech, ash). Will visit feeder for sunflower and thistle seeds; insects eaten in late spring.

curvirostra

Pine Siskin

Carduelis

As THE LEAVES OF FALL appear throughout the East, the crisp air carries the high-pitched *"sweeee – sweeee – ji–ji–jit – sweee"* of the first Pine Siskins. Often they will be seen in small groups bobbing along in the characteristic flight of the goldfinch family. They rise and fall in this undulating pattern, then quickly swing around and plummet downward, landing in the treetops.

The slim, dark bill separates them from the other groups of heavier billed finches. They use it to manipulate the catkins in trees until a shower of shucked parts filters from the tree. Then, as if a silent signal had been given, they explode from the treetops and are gone from sight.

The number of siskins reaching the southern edge of their range varies greatly. During some winters, thousands will be seen. Such invasion years often lead to residual populations that at times nest far south of their normal nesting range. In the western portion of North America, the mountainous areas and high plateaus host the species year-round.

ATTRACTING THEM TO YOUR GARDEN

Their favorite food is thistle seed and a thistle seed feeder will be emptied rapidly as a progression of birds occupy the pegs throughout the day. They are also attracted to trees with catkin-type fruit such as an alder or birch. Sunflower seeds are also taken.

DISTRIBUTION

Across Canada, northern U.S. and in the western mountains.

THE SONG

swee swee see jit jit jit see see

- **Note sequence:** Very reedy in quality and always incorporating the swee notes typical of the family.

- **Time of song:** All day.

- **Other birds with similar song:** Close to American Goldfinch (*Carduelis tristis*), see page 228, but more raspy.

IN BRIEF

- **Behavior:** Gregarious, relatively tame, often forms large flocks with other small finches.

- **ID:** Trim with dark brown streaking. Distinct yellowish wing bars with yellow also on sides of tail. Length 5 inches.

- **Habitat:** Coniferous woodlands and adjacent mixed woodland. In winter, backyards, gardens, parks.

- **Nest:** A shallow cup of grasses, rootlets, and mosses lined with feathers and fur.

- **Food:** Seeds and some insects. Thistle seeds preferred at feeder.

pinus   PINE SISKIN

Purple Finch

IF EVER THERE WAS A MISNOMER in identifying a bird by its name, this is one; "wine-colored" or "raspberry finch" would be much more fitting. The male is distinct in its wine-colored plumage that is washed with this color throughout. Note that there are none of the brown lines on the side that one sees in a House Finch. The female is more sparrow-like but has heavy streaking and a very distinct line over the eye contrasting to the deep chocolate-brown cap.

The song is a clear musical jumble of notes that invariably ends with the "to–eee" sound that from a distance sounds a bit like a towhee.

This bird has an interesting range in that it is a permanent resident on the East and West coasts with the whole population linking up via a wide band of summer birds across Canada. It is from this population that tremendous numbers migrate on their way to the southern United States in the fall. Their call note during migration is a distinct "tick" or "tick-tick." They have a wide range of habitat preference ranging from evergreen forests to mixed upland woods and parklands to suburban backyards.

ATTRACTING THEM TO YOUR GARDEN
During the winter months, this finch can be attracted by sunflower seeds at a feeder. It is very aggressive at the feeder site and will take control for hours, easily putting a dent in the day's seed supplies.

DISTRIBUTION
Across Canada down the West Coast and northeastern U.S.

THE SONG

sweet-ooooo sweet-ee-swee

- **Note sequence:** A bubbly series of rapid high-pitched whistles delivered in paired sequence, ending with a distinct tee-a or too-ee in a downward slur.

- **Time of song:** Morning and evening.

- **Other birds with similar song:** Similar to the House Finch (*Carpodacus mexicanus*), see page 226, but not as melodic or smooth.

IN BRIEF

- **Behavior:** Sings from treetops, moves about in groups; generally sedentary.

- **ID:** Often confused with the House Finch, the Purple Finch is larger, and the male is washed over most of its body with a rosy red color. The female is brown and heavily streaked. Length 6 inches.

- **Habitat:** Coniferous and mixed woodlands, parks.

- **Nest:** Shallow cup of twigs, rootlets, grasses, and bark in the crotch of a branch, usually of a conifer.

- **Food:** Principally seeds and fruit.

purpureus

Rusty Blackbird

OF ALL THE MEMBERS of the blackbird family that
inhabit North America, the Rusty is the most
northerly breeding species. Outside of a small area
in northern New England, Rusties appear in the
lower 48 states only during the migration period and
the winter. Habitat is an excellent key to locating
the Rusty Blackbird as it prefers wet swampy edges,
boggy areas, and brooklets. The thickets that
surround cattail marshes with their associated
tangles of alder are a prime location habitat.

The yellow iris color is a consistent feature for all
plumages. The color of the fall plumage gives the
bird its name, with the feather edges showing a rich,
coppery-rust color. Often the nape and back take
on a more solid copper hue.

In the spring and fall the song may be heard, a
raspy, creaking "*kish–a–lee*" often described as a
hinge badly in need of oil. They depart early in the
spring for their northern tree-line nesting grounds,
sometimes venturing into the bog areas of open
tundra with their stunted vegetation.

ATTRACTING THEM TO YOUR GARDEN

In migration or on wintering grounds they will visit
well-stocked bird feeders or take a variety of seeds
with sunflower and safflower being favorites.

DISTRIBUTION

Alaska through northern Canada,
winters southeastern U.S.

THE SONG

chack chack chack/kish-lay

- **Note sequence:** A short repeated
  rattle-like call; song is harsh and
  squeaky.

- **Time of song:** All day.

- **Other birds with similar song:**
  Like short, harsh Red-winged
  Blackbird (*Agelaius phoeniceus*), see
  page 238.

IN BRIEF

- **Behavior:** Walks about in wet areas
  turning over leaves for food.

- **ID:** The rust-colored tips on the
  back and wing feathers that give
  this bird its name appear only in the
  fall. During the breeding season, the
  male has a glossy back plumage,
  legs, and bill. Length 9 inches.

- **Habitat:** Almost always near water,
  boggy edges of northern coniferous
  woods.

- **Nest:** Bulky cup of grasses and
  twigs often in willow or alder.

- **Food:** Insects and their larvae,
  spiders, seeds, and berries. Will
  visit feeder for seeds with other
  blackbird flocks.

Agelaius phoeniceus

Red-winged Blackbird

Agelaius

A SPECIES FAMILIAR to almost everyone, birder or otherwise, across the United States. Perched atop a cattail in marshes from coast to coast, the familiar "*conk–ca–ree*" is the sound one most often associates with this habitat. It tends to have two basic territories. The nesting territory is defended with reckless abandon by the male. From its song perch with vermillion epaulets flashing, it pursues any stray male that crosses the invisible line of demarcation. And it has a lot to defend, because it is a polygamous species, with one male maintaining several nesting females. The other territory is a feeding area which is less strongly guarded against other intruding male birds.

A substantial portion of the population migrates south and this spells havoc for some southern roost areas. Populations in some states can exceed 11 million birds in a fairly small area and have caused health hazards — not to mention the noise factor. Dramatic steps have been taken to disturb such roosts but in general no great inroads have been made in population reduction.

ATTRACTING THEM TO YOUR GARDEN

In migration or on winter sites, they will visit the backyard often in large flocks to feed on seeds, especially sunflower. Bird baths are also visited.

DISTRIBUTION

Across North America.

THE SONG

conk-a ree onk-a ree

Note sequence: Loud, short and ringing with the ending drawn out.

Time of song: Morning and evening.

Other birds with similar song: Like a Tricolored Blackbird but more liquid in tone.

IN BRIEF

Behavior: Polygamous; forms massive feeding flocks in the fall and winter.

ID: The male's bright red shoulder patches with pale yellow edging make it easy to identify in flight, but it may not be as visible when perched. The female is brown above and heavily streaked on the underside. Length 8¾ inches.

Habitat: Cattail marshes, upland grassland, wood edge, dry fields.

Nest: A cup of grasses, plant fibers, and bark strips. Placed in grass tussock, woven into marsh vegetation, or in crotch of scrub trees.

Food: Insects, invertebrates, and a wide range of seeds. In cornfields damages newly formed cobs.

phoeniceus

Yellow-headed Blackbird

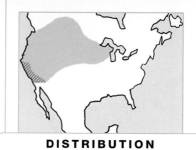

ALONG THE REEDY EDGES of freshwater lakes in western regions, the spring air is filled with squeaks and wheezes, much like the static on an old radio, produced by one of the handsomest of the blackbirds. Its name adequately describes its salient features. A large bird, all plumages from juvenile through female and first winter males show the same gold-yellow to the head and throat region. But it is the male in breeding plumage that has the full hood, with black wings that flash bold white rectangles in the wing coverts. A highly gregarious species throughout the year, the nesting marshes often harbor huge nesting colonies and the din created by the singing males can almost be overwhelming. Rather than foraging only in the marsh for food, they spread out over nearby agricultural land and mass in large flocks.

At the end of the breeding season they group with other blackbirds and concentrate in massive build-ups in agricultural lands and pastures to the south. Often these flocks consist only of winter-plumage males and young males, with the majority of females wintering farther south in Mexico.

ATTRACTING THEM TO YOUR GARDEN

In some instances the birds will join other blackbirds that come to feeders for seeds or any other form of handout. Cracked corn is taken readily as are sunflower seeds.

DISTRIBUTION

Western Canada and U.S.

THE SONG

chuck-chuck-zzzzzzzz eee ee

pip pop ee zee

- **Note sequence:** An electronic-sounding series of clucks and buzzing notes rising in intensity.

- **Time of song:** All day.

- **Other birds with similar song:** Has a few notes like a Red-winged blackbird (*Agelaius phoeniceus*), see page 238, but song is unique.

IN BRIEF

- **Behavior:** Has a slow, deliberate flight; very gregarious especially in the fall.

- **ID:** Black body, yellow head and breast, and white wing patches. The female is small and lacks the wing patch, but can be identified by its yellow breast streaked with white at the lower end. Length 9½ inches.

- **Habitat:** Freshwater reed and cattail marshes.

- **Nest:** Bulky cup of grass and sedge blades attached to reeds.

- **Food:** Attracted by seeds at feeder.

xanthocephal

Dolichonyx oryzivorus

Bobolink

THE BOBOLINK'S COLOR PATTERN appears to be upside-down in its orientation. Black underbelly contrasts sharply to the white of the shoulder, back, and rump and the rich yellow at the nape of the neck. In full song with head thrown back, the puffed yellow feathers of the nape make the head look exceptionally large. The song has been interpreted in many ways, from "*bob–o–link*" to "*link–link–Lincoln.*" Its metallic quality and jumbled notes, once heard, are easily remembered.

The nest, made of grasses, will be tucked well into a grass clump. The female lands and sneaks a significant distance to the nest. The eggs are reddish-brown splotched with a rich reddish purple.

In the fall, the males assume the same yellowish buff color of the females and head for Latin America. On fall mornings, thousands can be heard as they pass overhead, giving their distinctive "*pink–pink–pink*" call notes. In May, they will return to the rolling grasslands.

ATTRACTING THEM TO YOUR GARDEN

Not to be expected at a garden feeder as wide open fields are needed. Has shown up with Red-winged Blackbird flocks and then taken cracked corn.

DISTRIBUTION

Northern U.S. and southern Canada.

THE SONG

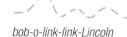

bob-o-link-link-Lincoln

Note sequence: A bubbly outpouring of metallic and warbled notes with a rolling pattern increasing in speed as it nears the end.

Time of song: All day.

Other birds with similar song: Unique and shouldn't be confused with any others.

IN BRIEF

Behavior: Flutters about on stiff wings calling.

ID: The breeding male has black underparts, a tan nape and white rump with white feathers at the base of its wings. The tail feathers of both male and female narrow to sharp points. Length 7 inches.

Habitat: Fields, meadows.

Nest: Of grasses tucked deep into the base of a grass clump.

Food: Seeds, insects.

oryzivorus

Sturnella magna

Eastern Meadowlark

Sturnella

No sound brings back birding memories of the fields and grasslands of North America as fast as the clear, two-parted song of the meadowlark. The sweet, liquid song is delivered most often from high up in a tree at the edge of a field. With head thrown back, the brilliance of the yellow chest catches the sunlight and appears to glow. The broad, black chest "V," in striking contrast to the yellow, is actually a disruptive color pattern that aids in the bird's camouflage. When seen from the back, the straw-brown and black streaking makes the bird virtually impossible to see in the grass.

Crossing a field, you may first see the birds jumping into flight close by, flashing the white edges of their outer tail feathers as they scale off on stiff wingbeats uttering a rapid, metallic "*bzzzeet*" series of notes. Though it may not look like a member of the blackbird family, when seen walking deliberately about, probing under objects with its stocky bill, one can see why it has been placed in that group.

The Eastern Meadowlark is stockier, lacks yellow on the feathers at the base of the lower mandible, and prefers a moister habitat than its look-alike western cousin of the dry grass plains.

ATTRACTING THEM TO YOUR GARDEN

Not expected at a feeder, but on rare occasions has come in with blackbird flocks and taken cracked corn.

DISTRIBUTION
Eastern two-thirds of U.S.

THE SONG

tee-tweedle ee do

- **Note sequence:** A loud series of high-pitched whistles and warbles.
- **Time of song:** All day.
- **Other birds with similar song:** Like Western Meadowlark but shorter in duration and less bubbly.

IN BRIEF

- **Behavior:** Calls from the top of a field-edge tree or plant tuft in open area; walks; deliberately flushes from close range; has a stiff-winged flight.
- **ID:** Yellow chest with broad black "V." Brown and black streaked back. Length 10 inches.
- **Habitat:** Grasslands, prairies, farmland.
- **Nest:** Cup of grasses tucked into a depression in field under a grass clump.
- **Food:** Wide range of insects, grass and weed seeds; in marshy areas also takes young snails.

magna

EASTERN MEADOWLARK

Baltimore Oriole

Icterus

STUDIES IN THE 1970s appeared to show that Baltimore or the western Bullock's Orioles were the same species as they interbred on the Great Plains river bottoms. However, their mixed gene pool selected for specific separate features and the Baltimore and Bullock's are no longer the combined Northern, but full species unto themselves.

One of the best known and most colorful of the blackbird family, not only does it bring us pleasure with its bright colors, its song is a mellow whistle delivered from the highest treetops before it darts off in a flash of color, giving off a series of loud "*check–check–check*" notes which gives away its blackbird background. In addition to enriching our lives with color and song, they also are most beneficial as they feed on destructive insects. These include the hairy caterpillars of serious pests such as the tent moth caterpillar, taken with relish by this species, but rejected by most other birds.

ATTRACTING THEM TO YOUR GARDEN

This is a species that has learned to live well with man. They enjoy the large shade trees that line the streets and are found in backyards. In winter, they will eat fruits and take suet from a feeder. Most move south.

DISTRIBUTION

Eastern half of N.A.

THE SONG

wee wee-too lee too/check-check check

- **Note sequence:** Introduced by clear whistles followed by a triplet of short whistled phrases; call is harsh and rapid.

- **Time of song:** Morning and evening.

- **Other birds with similar song:** Like the similar Bullock's Oriole but more liquid.

IN BRIEF

- **Behavior:** A bird of the high tree-tops. Very aggressive on territory.

- **ID:** Male is orange and black. Female can be distinguished by its deep yellow-orange breast. Two wing bars separate it from all but the Western Tanager. Length 8½ inches.

- **Habitat:** Open woodlands, riverine forest, shade trees in parks, yards, orchards, and gardens.

- **Nest:** Pendulant, placed at very tip of limb, well made of grasses and plant fibers.

- **Food:** Principally insects including hairy caterpillars, with some fruit and seeds; fruits taken in winter, when it will also take suet at feeder.

galbula   BALTIMORE ORIOLE

Orchard Oriole

Icterus

It always seems to come as a surprise to people when they learn that orioles are members of the blackbird family. Certainly they are among the most colorful of the group. The Orchard Oriole male has beautiful chestnut plumage at maturity with a black hood and cowl. In its first year, it is more the typical oriole in color, being a lemon-yellow with black face and bib. It is the smallest North American oriole and very trim in stature.

In many areas, this bird has become the oriole of tree-lined streets and parks and, like the other orioles, stream-side shaded groves. The Orchard Oriole is a gregarious species, nesting in loose colonies. The nest is a shallow cup of woven fibers in the fork of a limb, but not nearly as pendulant as a Baltimore Oriole's.

The song is made up of loud, clear, whistled notes, bubbling runs of cascading slurs, and metallic notes often ending with or introduced by a clear phrasing of *"what–cheer."*

ATTRACTING THEM TO YOUR GARDEN

This is a species that will adapt readily to a garden setting, and if twine or string is placed out in short strands from a hanging mesh basket, they will often come repeatedly to gather this material for nesting. Availability of water in a bath is another feature that will attract all types of oriole.

DISTRIBUTION

Eastern half of U.S.

THE SONG

what-cheer-kalee kalee-tolee

tolee-tolay what-cheer

- **Note sequence:** A rolling bubbly series introduced or ending with a distinctive what-cheer.

- **Time of song:** Morning and evening.

- **Other birds with similar song:** Similar to Baltimore Oriole *(Icterus galbula)*, see page 246, but more bubbly and sweet whistles.

IN BRIEF

- **Behavior:** Sociable, nests in loose colonies; often hangs head downward.

- **ID:** For the first year of its life, the male is a yellow-green like the female, but it then turns a deep chestnut and gains a jet-black hood. Length 7 inches.

- **Habitat:** Shade trees, orchards, riverine forest, open woodland, parks.

- **Nest:** A shallow cup of woven dried grasses and plant fibers suspended in fork of outer limb, usually of fruit tree.

- **Food:** Mainly insects, spiders, some berries, and seeds. Attracted to feeder by water.

spurius

ORCHARD ORIOLE

Useful contacts and addresses

American Birding Association
PO Box 6599
Colorado Springs, CO 80934
Telephone: 800-850-2473
Web site: www.americanbirding.org

American Ornithologists Union
Suite 402
1313 Doley Madison Blvd
McLean, VA 22101
Telephone 505-326-1579
Web site: www.aou.org

Cornell Laboratory of Ornithology
159 Sapsucker Woods Rd
Ithaca, NY 14850
Telephone: 800-843-2473
Web site: www.birds.cornell.edu

National Wildlife Federation
11100 Wildlife Center Drive
Reston, VA 20190-5362
Telephone: 800-822-9919
Web site: www.nwf.org

National Audubon Society
225 Varick Street
7th Floor
New York, NY 10014
Telephone: 212-979-3000
Web site: www.audubon.org

Nature Canada
85 Albert St
Suite 900
Ottawa, ON K1P 6A4
Telephone: 613-562-3447
Web site: www.naturecanada.ca

Birds Australia
Suite 2-05
60 Leicester Street
Carlton
VIC 3053
Australia
Telephone: +61 3 9347-0757
Web site: www.birdsaustralia.com.au

Royal Forest and Bird Protection Society
Level One
90 Ghuznee Street
PO Box 631
Wellington
New Zealand
Telephone: +64 4 385-7374
Web site: www.forestandbird.org.nz

British Ornithologists Union
BOU
PO Box 417
Peterborough
PE7 3FX
Telephone: +44 1733-844-820
Web site: www.bou.org.uk

Royal Society for the Protection of Birds
The Lodge
Potton Road
Sandy
Bedfordshire SG19 2DL
Telephone: +44 1767 680-551
Web site: www.rspb.org.uk

Wild Bird Society of Japan
1/F Odakyu Nishi Shinjuku Building
Hatsudai Shibuya-ku
Tokyo 151-0061
Japan
Web site: www.wbsj.org

Bombay Natural History Society
Hornbill House
Shaheed Bhagat Singh Rd
Mumbai 400001
India
Telephone: +91 22 2282-1811
Web site: www.bnhs.org

Index

Page numbers in *italic* refer to illustrations

Credits

Quarto would like to thank Mary Guthrie at the Macaulay Library of Cornell University's Laboratory of Ornithology.

Quarto would also like to thank the following for supplying photographs reproduced in this book:

Page 6 G. Ronald Austing/Photo Researchers, Inc.; 7 Ken Thomas/Photo Researchers, Inc.; 8 Thomas W. Martin/Photo Researchers, Inc; 9 John Bova/ Photo Researchers, Inc.; 13 Brock May/Photo Researchers, Inc.; 16 Jim Zipp/Photo Researchers, Inc.; 17 Doug Wechsler/Vireo; 19 David Weintraub/ Photo Researchers, Inc.; 20 Steve Maslowski/Photo Researchers, Inc.; 25 RollerFeeder available to buy from www.rollerfeeder.com; 27 G.Bailey/Vireo; 29 Phil Farnes/Photo Researchers, Inc.; 32 Ken Thomas/Photo Researchers, Inc.

The author would like to thank the following for their contributions:

My wife Carolyn and my two sons, Adam and Eric, who have always been so supportive of my birding endeavors.

Bill Martha for his work, past and present, on birdsongs for this book.

Dan Cinotti who has spent hours afield studying warblers with me.

Liz Pasfield for her considerable efforts and patience during this project and seeing it to fruition.